S0-CDP-420

The Lilies of China

中国百合花

The Lilies of China

THE GENERA

Lilium, Cardiocrinum, Nomocharis and *Notholirion*

STEPHEN G. HAW

with descriptions of the species by

LIANG SUNG-YUN

translated from

FLORA REIPUBLICAE POPULARIS SINICAE

TIMBER PRESS · PORTLAND, OREGON

First published 1986

First published in the USA by
Timber Press
9999 S.W. Wilshire
Portland, OR 97225
USA

ISBN 0-88192-034-7

Printed in Great Britain

Contents

Line drawings

List of illustrations

Introduction

This book covers the four genera which might be considered to be true lilies: *Lilium, Cardiocrinum, Nomocharis* and *Notholirion*. The word 'lily' is used in it as a general term for all the plants of these genera. They are so closely related that in the past almost all the species have been included within *Lilium*. Only a few *Nomocharis* species have never been placed in that genus, but the division between these two genera is very hard to draw, and has been altered several times by taxonomists.

Not long after I arrived in China in September 1981 I was able to obtain a copy of volume 14 of the *Flora Reipublicae Popularis Sinicae*, which includes descriptions of the genera *Lilium, Cardiocrinum, Nomocharis* and *Notholirion*. I immediately realized that the information in this work would be of great interest to western gardeners and botanists, very few of whom would be able to read the Chinese original. I have myself had an interest in lilies for almost as long as I can remember, and decided to begin work on a translation. This was completed during the course of the next two years, while I was studying and working in China at the University of Shandong in Jinan. Gradually the concept of this book as it now appears, in which the translation from which it began makes up only about a quarter of the whole text, took shape in my mind. After my return to Britain late in 1983 it was gradually transferred to paper.

During my two years' residence in China, and in the course of shorter visits both before and afterwards, I travelled as widely as was then possible in an attempt to find and photograph Chinese lilies in the wild. I was not able to visit the richest areas, in western Sichuan, north-west Yunnan and south-east Tibet, but nevertheless had some success. My extensive travels also enabled me to gain a clear impression of the vastness of China and of its geographic and climatic diversity. The richness of China's flora has, of course, long been appreciated by westerners with an interest in plants. It is the home of almost half of all lily species, with the greater part of the main centre of distribution of lilies within its borders. The information in the *Flora RPS* is correspondingly important. I have also made use of other Chinese sources, especially in the course of my researches into the history of lilies in China. There is a long continuous history of plant husbandry and horticulture in China, which has had an important influence on western gardens, though few westerners have any knowledge of this.

The transliteration of Chinese place names has given rise to much confusion, which has been compounded by the fact that there have been many name-changes during the past century. It is often extremely difficult to identify places referred to by western plant collectors who worked in China in the nineteenth and early twentieth centuries. Farrer's 'Siku', for example, is now Zhouqu. Virtually every collector seems to have used his own transliterations. Thus, the same place may be called 'Tsingki' by one collector, and 'Chingchi' by another. The present-day name is Hanyuan! I have tried to reduce confusion by using the standard Pinyin transliterations of place names published by the government of the People's Republic of China. Where another name has been commonly used in the past I have given it in brackets after at least the first occurrence of the modern equivalent. The Pinyin system is also used throughout this book whenever any Chinese word appears in transliteration. As this system may still be unfamiliar to many readers, its main differences from old systems and a guide to its pronunciation are given here.

Pinyin	*Pronunciation*	*Old equivalent(s)*
b	Like English b	p
p	Like English p, but more heavily aspirated	p'
d	Like English d	t
t	Like English t, but more heavily aspirated	t'
z	Like dz in English	ts
c	Like ts in English	ts'
zh	Like dj in English	ch
r	Like a combination of r with French j (a sound not found in any European language)	j
j	Like English j	ch, ts, k
q	Like English ch	ch', ts', k
x	Like English sh, but lighter and more aspirated	hs, sh, s, h
i	(After zh, ch, sh, r) Somewhat like English r (another sound not found in European languages)	ih, e

u	(After j, q, x, y) Like French u	ü, u
e	Usually like English a in 'ago'	o

This is by no means a complete guide, but covers the sounds most likely to cause problems. The pronunciations given can be no more than approximate, but should, it is hoped, be of some assistance.

The names of Chinese provinces are generally familiar to westerners in their old 'post office' transliterations. Their modern Pinyin versions are sometimes not very easily recognized. Here is a complete table of equivalents.

Old form	*New form*	*Old form*	*New form*
Heilungkiang	Heilongjiang	Sinkiang	Xinjiang
Kirin	Jilin	Tsinghai	Qinghai
Liaoning	Liaoning	Szechwan	Sichuan
Hopei	Hebei	Yunnan	Yunnan
Shantung	Shandong	Kweichow	Guizhou
Honan	Henan	Kwangsi	Guangxi
Kiangsu	Jiangsu	Kwangtung	Guangdong
Anhwei	Anhui	Fukien	Fujian
Hupei	Hubei	Chekiang	Zhejiang
Shansi	Shanxi	Kiangsi	Jiangxi
Shensi	Shaanxi	Hunan	Hunan
Kansu	Gansu	Taiwan	Taiwan
Ningsia	Ningxia		

I have, however, continued to use the familiar names 'Tibet' and 'Inner Mongolia', which are long-established in English usage, and are in any case not romanizations of Chinese names. 'Manchuria' is occasionally used in this book as a name for the whole of the extreme north-east of China, comprising the three provinces of Heilongjiang, Jilin and Liaoning.

I have had assistance during the writing of this book from many people and institutions. Thanks are especially due to Miss Liang Sung-yun of the Botanical Institute of Academia Sinica in Beijing, for her permission to publish my translation of her descriptions of the species in the *Flora RPS*, and also to Professor T. T. Yu of the same institute for his advice and assistance. The Royal Botanic Gardens, Kew, British Museum (Natural History) and Royal Botanic Garden, Edinburgh, kindly allowed me to consult the specimens in their herbaria, and I am most grateful to all their staff for the help which I received during this work. I would like to mention in particular Mr I. Hedge, Mrs V. A. Matthews and Dr D. F. Chamberlain of RBG Edinburgh, who gave up much of their valuable time to discuss lilies with me. I was also allowed to photograph in the garden at Edinburgh, where one *Nomocharis* species and two of *Notholirion* from collections made from the wild in China during the Sino-British Expedition to Cangshan in 1981 were flowering; my thanks are due to Mr R. McBeath, who is in charge of the peat garden there, and to his staff. The British Council awarded me the scholarship to study in China which took me to the University of Shandong, where this project began. I must also thank my family for their support, particularly my brother, Graham Haw, of Gateway Business Systems, who helped me to use my word-processor effectively. Last but not least, I must mention Ann Evans, my companion on long journeys to remote and often uncomfortable areas of China in search of wild lilies, who also suffered my unsociability during the writing of this book.

I have attempted to be thorough in my researches and as accurate as possible. Any errors in this work are entirely my own responsibility.

CHAPTER ONE

Chinese lilies in their natural environment

(i) Distribution

There are altogether more than 90 species in the genus *Lilium*, distributed throughout the north temperate zone. About a dozen of these grow in Europe (though half are restricted to the Caucasus, and only just merit being considered European); some two dozen occur in North America. The remaining 50-odd *Lilium* species, together with all the species of the closely related genera *Cardiocrinum*, *Notholirion* and *Nomocharis*, are Asian plants. Only one lily, *Lilium martagon*, has an area of distribution which extends across both Asia and Europe. In south and south-east Asia a few species carry the overall range of the genus southwards into the tropics. *Lilium philippinense* from northern Luzon (approximately 17–18° north) and *Lilium neilgherrense* (now considered to be a variety of *Lilium wallichianum*) from the Nilgiri, Pulney and Cardamon Hills of southern India (about 11° 30′ N) are the *Lilium* species of most southerly occurrence. In the other genera, *Notholirion campanulatum* is recorded from slightly further south, in the highlands of Sri Lanka.

China, with almost 40 species of *Lilium* and all but two or three of the species of the other genera, is home to more lilies than any other country in the world. They are, moreover, widely distributed throughout each of its provinces and regions. Only one large section of the country is devoid of lilies, the far north-western area comprising the great deserts of Xinjiang and the adjacent cold and arid regions of the Qinghai-Tibet plateau. *Lilium martagon* var. *pilosiusculum* does, however, extend into the mountains of northern Xinjiang (the Altai and Tarbagatai ranges), and two species reach eastern Qinghai. Quite a number of lilies grow in the Himalayan zone of south-east Tibet, extending as far north as the valleys of the Yarlung Zangbo (Yalu Tsangpo) and its tributaries and hills to the north of Lhasa (about 30° N). Further west and south, in the adjacent mountainous areas of Yunnan, Sichuan, Upper Burma and Arunachal Pradesh, the number of lily species increases rapidly. This eastern Himalayan and south-west Chinese region is, in fact, the major world centre of distribution of lilies (see map, Fig. 1).

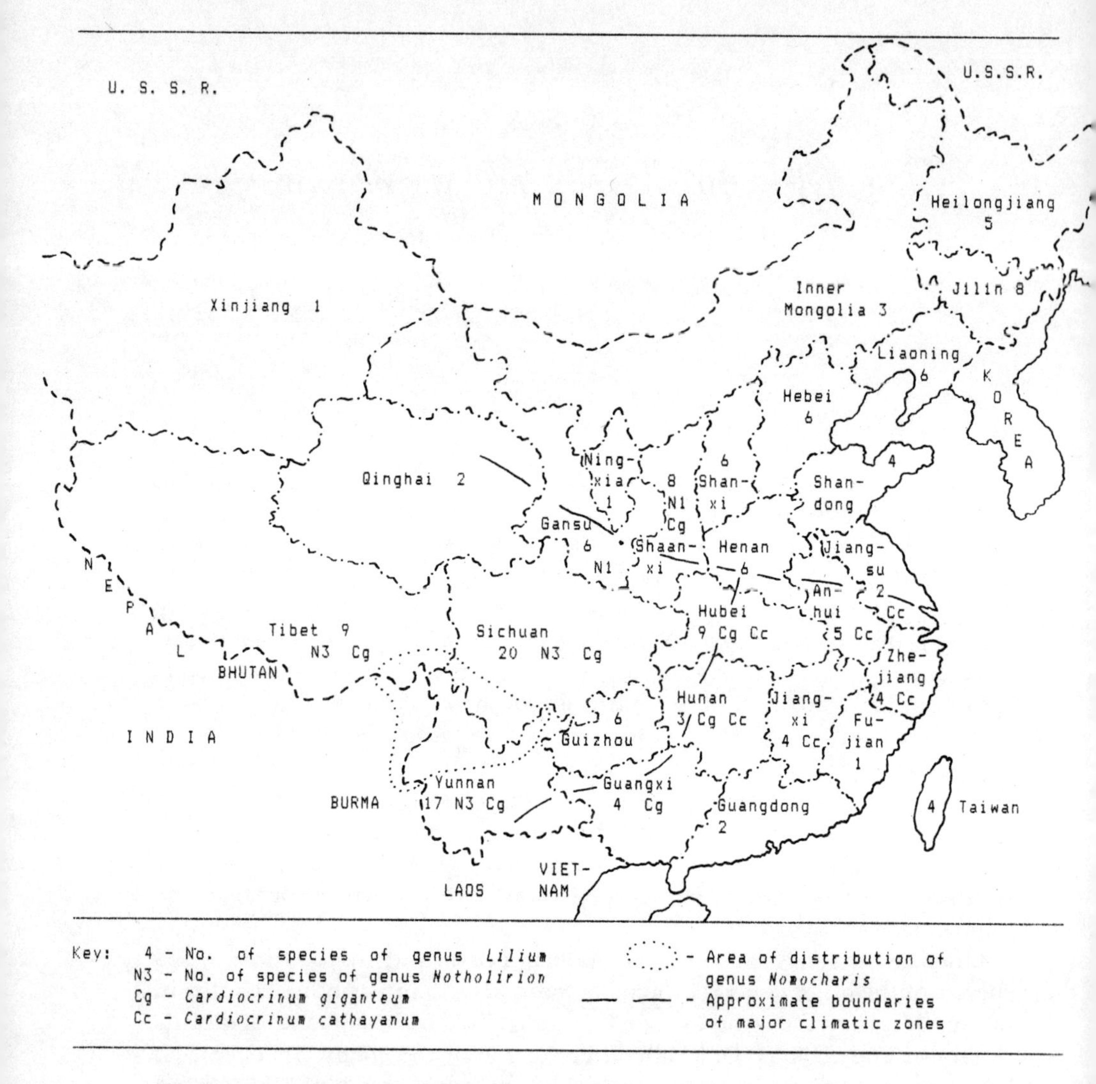

Fig. 1.
Map of China, showing numbers of Lily species per Province (Region).

Fig. 2.
Table showing the distribution by province of the Chinese species of Lilium.

Province (Region) / Lilium species	Xinjiang	Inner Mongolia	Heilongjiang	Jilin	Liaoning	Hebei	Shanxi	Shaanxi	Ningxia	Gansu	Qinghai	Henan	Anhui	Shandong	Jiangsu	Zhejiang	Jiangxi	Hunan	Hubei	Guizhou	Sichuan	Yunnan	Tibet	Guangxi	Guangdong	Fujian	Taiwan	TOTAL Provs.
concolor		X	X	X	X	X	X	X				X		X					X			X						11
dauricum		X	X	X	X	X																						5
brownii						X	X	X		X		X	X			X	X	X	X	X	X	X		X	X	X		16
regale																					X							1
formosanum																											X	1
longiflorum																											X	1
leucanthum										X									X		X							3
sulphureum																				X	X	X		X				4
sargentiae																					X							1
bakerianum																				X	X	X						3
nanum																					X	X	X					3
lophophorum																					X	X	X					3
souliei																					X	X	X					3
sempervivoideum																					X	X						2
amoenum																						X						1
paradoxum																							X					1
henrici																					X	X						2
speciosum													X			X	X	X						X			X	6
henryi																	X		X	X								3
rosthornii																			X	X	X							3
nepalense																				X	X	X	X					4
wardii																							X					1
stewartianum																						X						1
taliense																					X	X						2
duchartrei										X											X	X	X					4
davidii							X	X		X		X							X		X	X						7
leichtlinii			X	X	X	X	X	X																				6
lancifolium				X		X	X	X		X	X	X	X	X	X	X	X	X	X		X		X	X				17
pumilum		X	X	X	X	X	X	X	X	X	X	X		X														12
cernuum				X																								1
callosum			X	X	X							X	X		X	X			X						X		X	10
papilliferum								X													X	X						3
fargesii								X											X		X	X						4
xanthellum																					X							1
tsingtauense													X	X														2
distichum				X	X																							2
martagon	X																											1
TOTAL No. of species	1	3	5	8	6	6	6	8	1	6	2	6	5	4	2	4	4	3	9	6	20	17	9	4	2	1	4	

Sichuan, with a total of 28 species of the four genera, and Yunnan, with 27, have easily the greatest number of lilies of any of the provinces and regions. Together with Tibet, where 15 occur, these provinces have two-thirds of the Chinese species of lily and almost a third of the world total. From this centre of concentration, the number of lilies found in each province drops off extremely sharply towards the south-east, where Guangdong and Fujian provinces have only two between them, and rather less rapidly towards the north-east. The Qinling mountains of southern Gansu and Shaanxi support a moderate number of species, as does mountainous western Hubei. Further east, there are rather less lilies found in each province, but after falling to six in Shanxi, Hebei and Liaoning, there is an increase to eight in Jilin. There would seem to be another concentration of lily species in Japan and Korea (where altogether about a score are found), which extends across the Changbai Mountains into north-east China. All of the lilies found in Jilin also occur in Korea, and more than half are native to Japan. But only one or two of them grow also in the south-western provinces of China, so this secondary centre of distribution in Manchuria is clearly quite distinct from the main one.

The overall pattern of distribution outlined above indicates an east or south-east Asian origin for lilies. Not only do more than half the species of the four genera occur in China, Burma, Korea and Japan, but the three genera other than *Lilium* are more or less confined to this region. Of the three species of *Cardiocrinum*, one is Japanese, one is from east and central China, and one from west China and the Himalaya. The genus *Notholirion* is west Chinese and Himalayan in distribution, reaching westwards along the Himalayan range as far as Kashmir and Afghanistan, and with an interesting outlying station in Sri Lanka. The genus *Nomocharis* has a range which almost precisely coincides with the main Sino-Himalayan centre of distribution for lilies generally. *Nomocharis* seems to be an extreme development from the genus *Lilium*, while *Notholirion* and *Cardiocrinum* exhibit features which may be considered more primitive (such as the form of the bulb, which in these genera is still composed of the swollen bases of the leaves or leaf-stalks. In *Lilium* and *Nomocharis* basal leaves are almost always lacking in mature plants, occurring in only a couple of species). Additional indications of Asian origins are the southward extension of distribution in south and south-east Asia, and the great range of variation of form shown by the Asian lily species. Even after *Nomocharis*, *Cardiocrinum* and *Notholirium* have been excluded from *Lilium*, the latter genus remains extremely heteromorphic, and certainly exhibits its greatest diversity in eastern Asia. It is only necessary to consider the differences between, for example, the tall trumpet-lilies such as *Lilium brownii*, and the peculiar *Lilium paradoxum*, with its whorled leaves and bell-shaped flower; or the dimorphic leaves and strongly recurved perianth segments of *Lilium henryi*, and the small, starry-flowered *Lilium concolor*, to realize how great the variation is. The relationships between the different types of *Lilium* are complex and still poorly understood.

About half of the Chinese lily species occur only in China. Many are restricted to just a few provinces within China, and often to quite limited areas within those provinces. This is particularly true of those from the south-west. Almost a score of lilies are endemic to areas within western Sichuan, north-west Yunnan and south-east Tibet, and the figure would be quite considerably higher if those species which extend a short distance beyond national borders into Burma, India or Bhutan were added. Included among these lilies of very limited range are such little known rarities as *Lilium paradoxum* from Tibet, *Lilium stewartianum* from north-west Yunnan, and *Lilium amoenum* from central and southern districts of the same province. But even the familiar *Lilium regale*, which has adapted so well to a diversity of conditions in our gardens, has a very restricted area of distribution in the wild, growing only in the valley of the Min River in western Sichuan. The rugged topography of this region, with high mountain ranges separated by deep river valleys, and considerable local variation in climate, has undoubtedly led to the isolation of plant communities and to genetic divergence between the isolated communities. The wide variety of habitats available has also enabled the survival of a corresponding variety of plants.

There are also, however, a number of other lily species which have large ranges covering considerable areas of China. *Lilium lancifolium* is the most widespread, occurring in no less than 17 provinces (for reasons stated in Chapter 3, I do not believe that this wide range of distribution results from having been long in cultivation). It is closely followed by *Lilium brownii*, with *Lilium pumilum* also being found in a dozen provinces. *Lilium concolor* and *Lilium callosum* are other Chinese lilies with rather large distribution ranges. All these lilies except *Lilium brownii* also occur outside of China, in Japan, Korea or Soviet Asia. A few species, though they grow only within small areas of China, have extensive ranges beyond its borders. *Lilium martagon* is one such lily. The most wide-ranging of all lily species, it is found from Portugal across Europe to Mongolia and Siberia, and in China occurs in part of Xinjiang. *Lilium dauricum* is another example, with a range including parts of Japan, Korea and Mongolia, a considerable area of Soviet Asia, and five provinces in China's north-east. (Figure 2 tabulates the distribution, by province, of the Chinese *Lilium* species.)

Species of the genera *Cardiocrinum* and *Notholirion* are, in general, quite widely distributed. *Cardiocrinum cathayanum* spreads across central China to the east coastal provinces of Zhejiang and 'Jiangsu. Its larger cousin, *Cardiocrinum giganteum*, extends from west-central areas through Sichuan and Yunnan to the Tibetan Himalaya. Crossing the mountains, it reaches Burma, India, Bhutan and Nepal, to a western limit in Kashmir. The *Notholirion* species have similarly extensive ranges, with *Notholirion bulbuliferum* occurring from southern Shaanxi to eastern Nepal. The species of the genus *Nomocharis*, by contrast, are much more limited in their areas of distribution. *Nomocharis aperta* has the largest range, from the Muli region of Sichuan across north-west Yunnan to the extreme north of Burma, and northwards to the Tibetan border. This is scarcely 300 miles from east to west, and rather less from north to south. *Nomocharis basilissa* is known from only a small area on the frontier between Yunnan and Burma, between about 26° 30′ and 27° 30′ N. All the *Nomocharis* species are, in addition, confined within quite narrow altitudinal ranges.

(ii) Habitats

Poor understanding of the conditions under which lilies grow in the wild has probably contributed more than any other single cause to failure to manage them well in cultivation. There has been (and still often is) a tendency for gardeners to think that all plants that originate from China have similar cultural requirements, but this is quite erroneous. The natural environments of Chinese lilies vary very considerably, as ought to be expected in view of the large area of distribution and wide altitudinal range throughout which they are found. Climate, soil and other environmental factors can be very different indeed in one part of China from what they are in another.

There are, nevertheless, certain fundamental requirements which are common to the great majority of Chinese lilies. Most obviously, they are in general plants of hill and mountain slopes, often growing in stony or rocky places, and sometimes even on cliffs. There is no doubt that they demand good drainage, which has been found essential in their cultivation. Most of the species also commonly grow in thick scrub, thrusting their tall stems from heavy shade beneath the tangle of branches into the bright sunlight above. I well remember the difficulty with which I approached a tall flowering stem of *Lilium brownii* var. *viridulum* growing in the Lu Shan range in central China (Jiangxi province), forcing my way through dense, waist-high thickets among which scrambled spiny *Smilax* vines. But although they often grow among bushes, most Chinese lilies are not woodland plants; they enjoy exposure to full sunlight at least on the upper half of their stems. The major exceptions to this are the *Cardiocrinum* species, which normally grow in woods and forests with considerable overhead shade.

There is very little more that can be said about the general conditions under which Chinese lilies grow, without having to note far too many exceptions for the generalizations to be useful. It is, in fact, rare to find more than one species of lily occurring in exactly the same location in China. Therefore it is obvious that they must all have, at least to some extent, their own particular requirements. But sometimes several species may be found growing at no great distance from each other. In the Lu Shan mountains, for example, apart from *Lilium brownii,* which has already been mentioned above, both *Lilium speciosum* var. *gloriosoides* and *Lilium lancifolium* occur within an area of a few square kilometres. (*Cardiocrinum cathayanum* is also recorded from woodland on this same rather small block of mountains, of which the highest point is only 1474 metres above sea level.) Although they occupy rather different places within the range of habitats on Lu Shan, many of their growing conditions, such as climate, are of course common to all of them. It is therefore possible to group the natural habitats of Chinese lilies into a few broad divisions, as a convenient and useful way to consider their general aspects.

China east of the Tibetan plateau shows a marked difference in climate between north and south. The dividing line runs roughly along the Qinling mountains in the west and then follows the course of the Huai river eastwards to the coast. North of this line annual precipitation is low (about 80cm down to 25cm or even less), and falls almost entirely in summer. There is great variation in rainfall from year to year, and droughts are of frequent occurrence. Winters are very cold, as well as being dry, with temperatures often falling to minus 10 or 20 degrees Centigrade, or even lower in the coldest areas. Parts of Inner Mongolia, for example, have an average January temperature of minus 23° C, and even in the coastal province of Shandong temperatures may sometimes fall to below minus 25° C. Frosts occur in this northern region in at least four months of the year, and often in as many as six. South of the line rainfall is higher and more certain, exceeding 150cm per annum in much of the region. There is still basically a monsoon climate, as in the north, but winter precipitation is much less sparse. Although summer temperatures in north and south do not differ markedly (July averages generally fall between 25 and 30° C), southern winters are much less severe. Temperatures in much of the south rarely fall very far below minus 5° C, and large areas in the south-east of the region only occasionally suffer any frost at all.

Topography and the proximity of the Indian Ocean, to the south-west, and the Pacific Ocean, to the south-east, complicate climatic conditions in southern China. The north, especially if the large north-western area where no lilies are found is excluded, is climatically much more uniform. Very little land rises much above 2000 metres, and oceanic influence is relatively weak. Nine lily species, all belonging to the genus *Lilium*, are found in this area, or ten if northern Xinjiang is included as being climatically similar in the mountainous parts where *Lilium martagon* grows. All but two or three of these lilies occur in China only within this northern division, but all extend beyond the borders to various parts of Korea, Japan, Mongolia and far eastern USSR. The greatest concentration (eight species) is in Jilin province, where the Changbai Mountains on the Korean border provide a large area of thinly populated and uncultivated land where lilies can find suitable environments (there is now a vast nature reserve covering 190,000 hectares in this mountain range). These are plants of grassy mountain slopes, clearings in forests, the edges of woodland or scrub on rocky slopes. Their requirements for moisture during the growing season vary somewhat, but are generally moderate to quite low. The most common and widespread of these species are *Lilium pumilum* and *Lilium concolor*. The former I have seen growing on hot, dry, rocky hill-slopes near Beijing, among scrub of which a major component was *Vitex negundo*. In this situation it reached about 60cm in height, and carried as many as five flowers on each stem. The hills near Datong in the north of Shanxi province, just to the south of the Great Wall and the border with Inner Mongolia, are even more arid, and the vegetation of the hillsides is sparse and open. Small bushes of *Clematis fruticosa* about 40cm high grow dotted about on the slopes, with an occasional larger bush of *Rosa* or *Caragana*. Here and there occur patches of creeping *Thymus serpyllum* var. *asiaticus* and of a prostrate *Potentilla* species. *Lilium pumilum* as I saw it here justified the botanist who gave it its Latin epithet (who, it has now been established, was Delile rather than De Candolle), for it was no more than 20–25cm in height, bearing only a solitary flower on each stem. In neither of these situations was this lily particularly plentiful, individual plants being quite widely scattered, but they were of sufficiently frequent occurrence to be considered reasonably common.

Lilium concolor grows in much greater quantity on ridges to the north of the summit of Tai Shan in Shandong province (altitude 1532 metres). It is most frequent in open, grassy areas there, but also occurs scattered among low bushes of *Rhododendron micranthum* and *Spiraea pubescens*. Often there is very little soil covering the rock, and the bulbs of the lily are wedged into fissures. It may reach up to about 60cm in height, but is usually smaller, with one to four flowers to a stem. In the Lao Shan hills near Qingdao on the south-east coast of the same province, I found it to be much less numerous. It grows there on grassy banks at the edges of fields at just a few hundred metres above sea level. I was only able to discover small, single-flowered specimens, with perianth segments spotted with black on their basal halves. Chinese botanists have assigned this form to the variety *pulchellum*, though in the west it has usually been given separate varietal rank.

Neither of these two lily species seems to show any preference for either alkaline or acid soils. Tai Shan, for example, is granitic, while Lao Shan is composed of a limestone which is soluble enough for there to be mineral water springs in the area. The hills west of Beijing where I found *Lilium pumilum* are mainly of limestone, but those near Datong were formed of a soft, crumbling sandstone. This tolerance of varying amounts of lime in the soil has been confirmed by experience in cultivation. Indeed, the majority of Chinese lilies are tolerant of variation in soil acidity. The only species so far known to have really strong preferences are *Lilium henryi*, which shows a definite liking for lime, and *Lilium speciosum*, which prefers an acid soil. Slight acidity is usually preferable for the others.

There are two of the north Chinese lilies which favour rather different habitats to the other plants of this group. *Lilium distichum* and *Lilium tsingtauense* apparently have a greater requirement for shade and moisture, and grow in light woodland or on the banks of streams. Both these species are, in fact, of local occurrence in China, and *Lilium tsingtauense* in particular now seems to be quite rare. They both occur in Korea, however, and *Lilium distichum* also has a considerable range in the USSR.

Many more lilies are found in the southern half of China, and their habitats are much more diverse. In the south-west, there are very high mountains and deep valleys, and altitude is therefore an important consideration. Whereas in north China the same lily species may grow from just above sea level to near the summits of the mountains (as I have personally observed with *Lilium concolor*), on the much higher ranges of the south-west this is not the case. There are very distinct altitudinal zones of vegetation in this region, and individual species of lily are usually more or less restricted to only one of these zones. They may thus be grouped in accordance with the heights at which they occur. There seem to be three major divisions of altitudinal range. The first is from sea level up to about 2000 metres; the second, from 2000 to 3000 metres; and the third, from 3000 metres upwards. These are naturally rather rough divisions, as the precise altitude limits of the vegetation zones vary somewhat according to latitude and local climatic conditions, but they seem to be accurate enough to be useful. The great majority of the lilies of south-west China are more or less confined to one or other of these ranges of altitude, with only about half-a-dozen species regularly occurring within more than one.

The climate at low altitudes is warm temperate to sub-tropical, and generally quite wet. Considering that winters are usually rather mild in this zone, the lilies which occur within it are surprisingly hardy. This group in fact includes many of the Chinese lilies that are reliable garden plants in Europe and America, such as *Lilium regale* and *Lilium henryi*. It seems particularly strange that *Lilium regale*, a plant of very limited distribution in the wild, should have taken so well to cultivation outside its native land. The bulb is, of course, protected throughout the winter by its covering of soil, and no doubt the comparative mildness and dampness of winters in this zone, with occasional quite severe freezes, demand considerable adaptability. Certainly such conditions are not dissimilar to those of the damp and uncertain British winters, which are apparently more difficult for plants to cope with than unremitting severe cold (judging by the number of perfectly frost-hardy plants that regularly perish during the winter in Britain). The lilies of this group are nevertheless susceptible to damage to their young shoots by late spring frosts, and benefit from being protected in some way. Planting among shrubs can provide such protection. In addition to the species already mentioned, this group includes most of the other lilies of section Regalia, plus some others such as *Lilium fargesii* and *Lilium lancifolium* (though this last lily is by no means restricted to south-west China). *Lilium davidii* and *Cardiocrinum giganteum* are other lilies which occur in this zone, though they also extend to higher altitudes. The *Cardiocrinum*, as a plant of woodland shade, is not typical in its growing requirements.

The most difficult group of lilies to cultivate are those from the moderate range of altitude between 2000 and 3000 metres. Some, like *Lilium amoenum*, have been introduced to the West but quickly lost to cultivation. Others, such as *Lilium bakerianum* and *Lilium nepalense*, have persisted (reinforced by new introductions) as challenges to the most experienced growers. They are often considered to be doubtfully hardy. This may be true of some forms, collected from low altitudes at the southern limits of their ranges, yet frosts are common throughout most of their areas of occurrence. It is much more probable that it is dampness in winter that is their main undoing, for although the monsoon thoroughly soaks them in summer, winters can be very dry indeed. Frost and protection by evergreen woodland, the roots of which remove moisture from unfrozen soil, reinforce the effects of climate. These lilies most often grow among shrubs at the edges of the woodland which is predominant in this altitudinal zone (except where it has been destroyed by man). The species that are in cultivation have occasionally been grown successfully out-of-doors in Britain, when it has been possible to meet their exacting requirement to be kept very dry in winter but given copious summer moisture. Most growers have found it easier to keep them alive under the protection of a cool glasshouse, however. The only lily of this group which seems to have settled even moderately well into cultivation is the beautiful *Lilium wardii*, but even this species tends to be short-lived, and is highly susceptible to virus diseases.

The remaining group of lilies from the south-west are those which grow at the highest altitudes, the sub-alpine to alpine zone. All the species of *Nomocharis* and *Notholirion* belong here. So do a number of *Lilium* species, most of which (belonging to section Lophophorum) are closely related to the genus *Nomocharis*. So closely related, in fact, that several of them were at one time included in that genus. Few of these alpine *Lilium* species are now in cultivation, though it is hard to be certain of how difficult to grow they really are, as they have only ever been available in small quantities from a few collections. Some of them, such as *Lilium paradoxum* and *Lilium sempervivoideum*, have never been in cultivation at all. Some appeared transiently, but soon dwindled away, and due to difficulties of access to their Chinese mountain homes have never been collected again. *Lilium souliei* and *Lilium lophophorum* are among those which were grown and even flowered (at the Royal Botanic Garden, Edinburgh, and on Vancouver Island in Canada), but then expired without producing offspring.

Lilium nanum has a wider range in the wild, including parts of the Himalayan range that have been more easily accessible in recent years than has south-west China. It has therefore been introduced more often and in greater quantity, and has persisted quite well from at least some of these introductions. None of these alpine lilies could be considered easy to grow, however. The conditions of their natural environment are difficult to reproduce, especially in areas of mild winters, such as the south of England. The climate of the high-altitude regions where they grow wild is extreme, both in the range of temperatures experienced and in quantities and annual distribution of rainfall. The monsoon brings several months of heavy, daily downpours to the mountains every summer, from about late May until early

October, while the rest of the year is often almost completely dry (the eastern Himalaya usually have more snow on them in summer than they do in winter, because of this pattern of precipitation). The daily range of temperatures during sunny periods of the year can be enormous, going from far below freezing at night to the twenties or even thirties Centigrade in the hottest part of the day, and higher in sheltered spots in direct sunshine. But if there is no sun in winter long periods of severe frost result. In the lower part of their altitudinal ranges, the lilies grow among scrub of dwarf rhododendrons, *Potentilla fruticosa* and other small bushes, in soil composed of grit mixed with large amounts of humus. At higher levels there are no woody plants, and the ground is usually more rocky, covered by a thin layer of turf. Amounts of available water are copious, both from rainfall and snow-melt, but drainage is extremely rapid. In cultivation they have generally succeeded best in areas of cold winters and at least moderate rainfall, such as northern parts of Britain. *Nomocharis* and *Notholirion* species have shown similar preferences, and have been most successfully grown in a few gardens in central Scotland. The high summer monsoon rainfall and cold, dry winters of their native mountains are hard to imitate in southern Britain.

In the south-eastern part of China only some half-dozen lilies occur, and many of these are widespread species also found in the north or west (*LL. brownii, lancifolium* and *callosum*). *Lilium formosanum* and *Lilium longiflorum* grow wild on Taiwan (though the latter species may not be native there), but are not found on the mainland except in cultivation. Except for high-altitude forms of *Lilium formosanum,* they are not reliably hardy in Britain. The south-eastern region is similar in climate to the low-altitude zone of the south-west, but has rather warmer winters because of the influence of the Pacific Ocean. *Lilium speciosum* var. *gloriosoides* is the outstanding lily of this area. I have seen it growing on the fringes of woodland at about 1200 metres altitude in the Lu Shan range in Jiangxi, the type location (it also grows in a few other provinces, including Taiwan). It is a very beautiful plant, and has indeed been considered the finest form of *Lilium speciosum* by many authorities including E. H. Wilson. It has strongly recurved, wavy-edged perianth segments of a brighter shade of red than most of its cousins, but unfortunately has proved rather short-lived in cultivation, with a fatal susceptibility to virus infections.

The conditions under which lilies grow in nature in China thus show considerable variation, and this must be taken into account when attempting to grow them in gardens. Generally speaking, the easiest to manage in normal garden conditions are those from north China or from low altitudes in the south-west. The alpine species may be persuaded to flourish in some gardens, particularly in such areas as Scotland and the north of England, in well-drained but moisture-retentive soils such as could be provided in a raised peat-bed. Recent collections of *Nomocharis* and *Notholirion* species from the Cang Shan range in Yunnan (made during the joint Sino-British Expedition of 1981) have been very successfully cultivated in the peat beds at the Royal Botanic Garden, Edinburgh. The most difficult lilies to cultivate are those from moderate altitudes in the south-west, which are extremely intolerant of winter wet and have only rarely been coaxed into flowering outside the cool glasshouse. The species that will succeed if a little care is taken to suit their needs are nonetheless quite numerous. Lilies are such beautiful plants that any trouble taken must be worthwhile.

(iii) Physiology

All lilies are bulbous plants. Their bulbs have no dry outer tunic (except in the genera *Cardiocrinum* and *Notholirion*), and are composed of fleshy scales, which may be few to numerous, and tightly to rather loosely imbricated. Sometimes the scales are jointed, as in *Lilium distichum*, but this condition is found in only a few of the Chinese lilies, while it is much more common with the North American species. *Nomocharis* and *Lilium* bulbs are essentially identical, but in the genera *Notholirion* and *Cardiocrinum* they are formed of the swollen bases of the leaves or leaf-stalks. Each year as new growth takes place, fresh scales develop from the heart of the bulb while the outermost ones tend to wither. In *Cardiocrinum* and *Notholirion*, flowering exhausts the bulb, and a mass of small bulbils, rather than a single large bulb, are left to grow on again to flowering size. This may take several years. Other lilies also form underground bulbils, as offsets either from the main bulb or from the stem. In a number of species (e.g. *LL. nepalense, wardii* and *duchartrei*) the stem is stoloniform, creeping obliquely through the soil and rooting before emerging above ground to produce leaves and flowers. Small bulbs are formed along these underground stems, sometimes at some distance from the parent bulb. Many other Chinese lilies, though not stoloniferous, send out roots from the stem above the bulb and produce bulblets among these roots. A few species even form bulbils in leaf axils of the aerial portion of the stem (e.g. *Lilium lancifolium* and *Lilium sargentiae*). In certain North American lily species, the bulbs are rhizomatous, creeping and even forming mats of scales.

Although lily bulbs spend part of each year in a dormant state, they do not like to be dried out like tulip bulbs, and may not be totally dormant for very long. Most lilies begin to put out new roots in the autumn, quite soon after the flowering stem has withered, in preparation for the following year's growth. The common practice of selling lilies in autumn as more or less dry bulbs is obviously, therefore, a very bad one, which only a few species are robust enough to survive (though some of the hybrids manage to cope with it, if not particularly well). Dried-out lily bulbs may sometimes appear to grow strongly enough in the season after planting, but having been unable to establish a sound root system sufficiently early, they fail to produce a flowering-size bulb for the following year. It must be remembered that the flowering of bulbous plants depends very largely on the previous season's growth. If the growing cycle of the plant is disrupted, the bulb formed at the end of the season may be smaller than that of the year before. Recently, the practice of selling bulbs of such plants as snowdrops 'in the green' has developed, as it was recognized that they recovered badly from desiccation. It is very much to be hoped that suppliers will also begin an equivalent practice with lilies, which would enable them to become more readily established in gardens.

Having produced some roots in autumn, the lily bulb will usually cease growth during the winter period. In its natural habitat, winter is often too cold and dry to allow continuing development. Growth recommences in spring, with further root activity and the emergence of the stem, which elongates from the centre of the bulb. Leaves grow along much of the length of the stem, and in most lilies are scattered in arrangement. Several *Nomocharis* species, as well as *Lilium* species of the section Martagon, most North American lilies and the anomalous *Lilium paradoxum*, have their leaves arranged in whorls, of which there may be several or just one on the stem. *Cardiocrinum cathayanum* is interesting in having most of the leaves on its stem clustered into a pseudo-whorl just below the middle, with a few additional ones scattered above.

As the shoot emerges from the bulb, the flower-buds are already starting to form. The growing period from its beginning in spring until the flowers open is usually quite long, however, and may be as much as five or six months in some late-flowering species. Most lilies flower during the summer months from June to September, on rather tall stems which raise their blooms above the level of surrounding vegetation. The flowers vary considerably in arrangement, forming a raceme, umbel or corymb, and being anything from erect to horizontal or downward-facing. The shape of the individual flowers is just as variable. Forms include the widely open saucer-shape of most of the *Nomocharis* species, the quite narrow bell of, for example, *Lilium nanum*, and the long trumpets of *Lilium formosanum* and its relatives. In many *Lilium* species the perianth segments are strongly recurved or revolute, as in *Lilium henryi* and the various groups of 'martagon' lilies. The flowers are pollinated by flying insects, the stigma commonly being receptive before the anthers have released their pollen, to ensure cross-fertilization. In many species, each plant is also self-sterile, and will produce little or no seed as a result of self-pollination. The seed capsule is globular, barrel-shaped or cylindric, and dries out as it ripens and then splits open to allow the seeds to scatter. Individual seeds are flattened and usually narrowly winged to assist in their dispersal by wind.

Lily seeds may germinate in one of several distinct ways. Species fall into two main groups according to their manner of germination. In one, the seeds when they sprout produce a seed-leaf above the surface of the soil. This is usually long and thread-like and often carries the seed-coat up out of the ground on its tip. The first true leaf appears shortly afterwards, while roots and a bulblet grow underground. This is known as epigeal germination. Lilies of the second group never produce a seed-leaf at all, but germinate by first producing a root below ground. This then develops a tiny bulblet before sending up the first leaf, which is not a cotyledon. This type of germination is called hypogeal. Most of the Chinese species of *Lilium*, and all the species of the other three genera, germinate epigeally. Only a few, such as *Lilium tsingtauense* and *Lilium speciosum*, have germination of the hypogeal type.

The seeds of many species which have epigeal germination sprout in the spring following their ripening and dispersal (immediate epigeal germination). Growth may be very rapid after sprouting has occurred, and some species, such as *Lilium formosanum*, have been known to come to flowering in one year from seed. A number of other species germinate in a similar way, except that the seed-leaves may take much longer to appear, often sprouting at irregular intervals over a period of more than two years (delayed epigeal germination). The seeds of lilies which usually germinate immediately will often also show this sort of delay, if they do not encounter suitable conditions for growth soon after they have ripened. They may then go into a dormant state, which will need to be broken in some way, usually by a sequence of warm and cold periods.

The seeds of those lilies which germinate hypogeally frequently produce roots and a bulblet soon after they are scattered, often in the autumn following flowering. Dormancy may also occur in species of this group, which requires varying sequences of warm and cold periods to be broken. After formation of the bulblet there is usually a period of rest, to correspond with the first winter. True leaves then grow in the spring, and are the first above-ground appearance of the plant. There is no subsequent difference in the fundamental growth patterns of the species of the different germination groups.

Immature lilies which have not reached flowering size produce basal leaves direct from the bulb. These are of similar shape to the stem leaves of the mature form of the species, but have a distinct stalk (or petiole), which is found on the stem leaves of rather few lilies. As bulbs grow larger, they begin to send up leafy stems, which at first do not carry any flower buds. After a varying number of seasons' growth, depending on the species and the growing conditions, the stem produced from the bulb will bear flowers. In the genera *Lilium* and *Nomocharis*, mature plants have no basal leaves (except in *LL. candidum* and *catesbaei*), but they normally occur in all the species of *Notholirion* and *Cardiocrinum*.

The life cycle of the lily is now complete. Seeds are produced from the flowers, and more new plants then grow from the seeds. Such is the natural sequence of growth, by which wild populations sustain themselves.

CHAPTER TWO

The propagation and cultivation of Chinese lilies

(i) Propagation by seed

It is appropriate to begin by considering the growing of lilies from seed, as this is usually the cheapest and easiest way to obtain lily species. It is also the best way to propagate clean stock, as virus diseases are apparently not transmitted to the seeds even if the parent plant is infected. Every grower of lilies should therefore sow seeds regularly, to ensure that new and uninfected plants are available should virus strike. This is also usually the easiest means to produce large quantities of new plants, and is the only practical way for the amateur gardener to propagate those lilies, such as the *Nomocharis* species, which rarely or never produce natural offsets, and are too delicate and precious for scaling to be regularly used. For all these reasons, the sowing of lily seeds is much to be encouraged.

The only serious drawback to this method of propagation is that it can be rather slow. Many lily species frequently take as long as five years to reach flowering size from seed. This period may be shortened, however, if optimum growing conditions can be provided. Most species will then flower after three seasons, and some can even be brought to flowering in just one year from sowing. But patience is a virtue much fostered by gardening, and the flowering of lilies would be more than adequate reward for even a decade of waiting.

Obtaining seed of many of the Chinese lilies is not always easy. Some species, of course, are simply not in cultivation at all, and await future introduction or re-introduction (probably in the form of seeds). Others are rare, or rarely set seed. Nevertheless, quite a large number are regularly available through the seed distributions of specialist societies, and anyone seriously interested in growing lilies should certainly join one of these societies and participate in the distributions (see Appendix 2). Chinese lilies generally do not hybridize very freely, so seed obtained is likely to come true (though there is no guarantee that it will be correctly named!).

Having obtained the seeds, it is worth giving some thought to their treatment. It has already been said that lilies do not all germinate in the same way, and this has to be taken into account when their seeds are being sown. In particular, it should be borne in mind that germination may not occur for a long time, occasionally as long as two or three years from sowing. The compost used, therefore, should be capable of remaining in good condition for periods of this length, and pots containing lily seeds should not be thrown away too soon! Even seeds of lilies which usually have immediate germination may become dormant and therefore fail to sprout quickly if they are not fresh enough when sown.

The compost used should be very free-draining, but capable of retaining a moderate amount of moisture. It is also best for the great majority of species if it is slightly acid (only *Lilium henryi* of the Chinese lilies now in cultivation definitely prefers an alkaline soil). I generally use a mixture of about two parts of a commercial John Innes seed compost with one part of peat and one part of sharp sand. More peat may be added for the alpine lilies, including the *Nomocharis* and *Notholirion* species, and more sand or grit for the species from north China. Fine leaf-mould may be used mixed with or instead of the peat, and is probably preferable for the woodland lilies such as *Lilium distichum* and the *Cardiocrinum* species.

It is not a good idea to sow lily seeds in shallow trays. It is difficult to provide adequate drainage in trays, which tend to waterlog (and dry out) much too easily. This not only has directly adverse effects on the lily seeds and seedlings, but also causes deterioration of the compost. This is obviously undesirable where germination may not occur until a long time after sowing. Even if they germinate quickly, the seedlings are best left to grow on undisturbed for at least one season and preferably two, as the tiny roots and bulbs of very young plants are easily damaged in handling. So it is also important that there should be room for the seedlings to grow and develop, which a tray will not provide. Ordinary plant-pots are much more suitable. I like to place about two inches of drainage (potsherds, stones and grit) in the bottom of the pots, before putting in the compost mixture. Plenty of sharp grit at the bottom not only allows good drainage, but also helps to prevent the entry through the drainage holes of pests and worms. Worms should be kept out of pots as far as is possible, as they disturb the seeds and seedlings and may ruin the texture of the compost. Plastic pots require less frequent watering, and I personally prefer them, but earthenware pots may also be used and will require less drainage.

Lily seeds are flat and roughly triangular, and when sowing them it encourages good germination if they are placed on edge rather than left lying flat. This is hard work and unnecessary where plenty of fresh seed is available, but worthwhile in other situations. The seeds are large enough to be quite easily handled individually, and should be well spaced apart when sown. After they have been placed or scattered on the surface of the compost, they should be covered completely with a further thin layer of fine compost, or with a thicker layer of sand or grit. Weeds are less likely to be a problem, and are more easily removed, if sand or grit is used.

The seeds of lilies which germinate epigeally are best sown early in spring. If sown in autumn as soon as ripe, they may sprout before winter sets in, and then have to try to survive the adverse winter conditions before they have had time to make much growth. Spring sowing prevents them receiving such a check. Seed of all the hardy species should be left exposed to frost, as this helps to break dormancy. The cotyledons ought then to appear within a few weeks of sowing, or as soon as the spring weather has become sufficiently warm. If germination is not immediate, it may not occur till spring of the following year, or even two years later. Seed of the *Cardiocrinum* species seems especially prone to delays of this kind.

Seed with hypogeal germination is a little less easily managed, if only because it may be hard to obtain it at the right time. If possible, it should be sown early in the autumn as soon as it is ripe, when the weather is still sufficiently warm for the formation of bulblets to take place before the onset of winter. Growth will cease during the winter cold, but leaves should appear the following spring. If not sown till spring, bulb formation may occur more or less immediately, but the appearance of leaves will usually be delayed for a whole season. This problem can be overcome by artificially providing alternation of hot and cold conditions. Seed should be placed in a polythene bag with a mixture of peat and grit, moistened (but not soaked!), and dusted with fungicide. The bag should then be sealed (though some ventilation is desirable) and placed in a warm cupboard until the formation of bulblets has occurred. It should then be transferred to the bottom of a domestic refrigerator for about three months to provide a period of dormancy, before the tiny bulblets are separated out and planted into pots. This is a fiddly process, and fortunately very few Chinese lilies need to be treated in this way. Only the species of section Martagon and *Lilium speciosum* have delayed hypogeal germination. The few other Chinese lilies which germinate hypogeally do not need the cold period of dormancy, and may be sown in spring and treated like those with epigeal germination.

Once the seedlings have appeared above the surface of the compost, they must be jealously guarded against slugs and snails, which will devour their leaves down to ground level if allowed access to the pots. After a few weeks, when they have established a good root system and are growing strongly, an occasional feed with a weak liquid fertilizer will help to build up the bulbs quickly. Feeding is especially useful during the second season of growth. When the young lilies have grown to sufficient size, they may be either potted on into larger containers or planted out into their final positions in the garden, where they must be tended carefully until they flower.

(ii) Propagation by other methods

The easiest and quickest way to propagate lilies is by means of naturally produced bulblets and bulbils. Nearly all lilies produce offset bulblets underground in one way or another, and a few species also produce aerial bulbils in leaf-axils on the stem. However they arise, these small bulbs may be grown on to produce new plants. Underground bulblets may form either alongside the original bulb or from the portion of the stem which grows below the soil surface. Most Chinese lilies send out roots from the stem above the bulb, and in many the stem grows obliquely, wandering through the soil and rooting in several places, before rising erect through the surface. Wherever the stem puts out roots, it may also produce bulblets. These can simply be left where they are, to grow on naturally to flowering size, but often, especially when they form directly above the parent bulb, it is better to move them to a new position. This should be done in autumn, as soon as the stem has died down, and care should be taken to damage their roots as little as possible. They should then have time to settle into their new site and become well-rooted before the cold of winter stops their growth.

Aerial bulbils are regularly produced in the upper leaf axils of *Lilium lancifolium*, *Lilium sargentiae* and *Lilium sulphureum*, and may occasionally be formed by some other lilies. They should be removed as early as possible, or as soon as they have reached the size of a small pea, then potted up and grown on in the same way as seedlings. If potted early enough, they will produce a good root system before the onset of winter, and will then be able to grow away strongly during the following season. After a year or two in pots, they should be quite large enough to be planted out in the garden.

The last important method of vegetative propagation of lilies is by scaling of mature bulbs. This involves lifting the bulbs, and either dividing them completely into their component scales, or removing only some of the outer scales and replanting their diminished hearts. Scales should be removed carefully and cleanly, to reduce the risk of leaving nasty wounds through which disease may gain entry. The scales thus obtained should then be scattered on the surface of compost in a pot, in much the same way as seeds, except that it is best to cover them at first with a layer of damp sphagnum moss. New bulblets should form at the base of each scale. These will send out roots and leaves just like any other bulblets. From time to time the moss should be lifted so that the progress of bulb formation can be checked. Once the bulblets are well-formed and have begun to put out roots, the sphagnum moss should be discarded and replaced with a layer of compost. It is common for two or three little bulbs to form on each scale, so they should be separated at this stage if necessary. When new bulbs are bought in a more or less dry state in late autumn, it is a good idea to take off a few of their outer scales and propagate new bulblets from them. Then if the parent bulbs fail to establish, as is often the case when they are acquired half-dry and with few healthy roots, there will at least be some young plants coming on to replace them. Small bulbs will in any case usually settle into a new environment better than mature ones.

(iii) Cultivation

Having acquired lilies, by whatever means, it remains to grow them successfully. Lilies are not among those plants that will flourish in virtually any situation. It is worthwhile to give careful consideration to their requirements before they are actually planted. Some preparation of the site will almost certainly be necessary.

The essential prerequisite for success in growing lilies is to be able to provide them with adequate drainage. Whatever other preparations may be made, if the soil is waterlogged in winter the bulbs will certainly rot. A sloping site from which water runs off easily is undoubtedly preferable, but if the soil is naturally fast-draining this may not be necessary. On heavier soils the incorporation of plenty of grit will allow water to drain away more easily, but it must be ensured that the water does not merely lie a little further below the surface. Land drains must be used if this is likely to be the case. Raised beds can be an excellent solution to drainage problems, and a number of Chinese lily species will grow very well in raised rock or peat gardens. The roots of shrubs and trees also assist drainage by removing large quantities of water from the soil, and many lilies flourish when planted among shrubs. It must be remembered, however, that it is in winter that drainage is most important, and deciduous shrubs take up little or no water at that season.

It is often said that lily bulbs should be planted deeply. The stem-rooting kinds certainly need sufficient depth to be able to send out plenty of roots above the bulb, but it is not wise to plant too deeply. Any problems with drainage will become worse as depth increases, and many bulbs are in any case able to pull themselves deeper into the soil with contractile roots if they are at first too shallowly planted. Young lily bulbs will certainly do this, and should be placed at a depth of no more than 10–12cm (3–4in). Large mature bulbs may be set deeper, but it is rarely necessary to cover them with more than twice their own height of soil. They should be planted more deeply in light soils than in heavy ones. Of the Chinese lilies, only the *Cardiocrinum* species need to be planted shallowly, with the bulb only just covered by leafy soil.

The best time for moving and planting lilies is early autumn, as soon as the stems have withered. When lifting established bulbs, great care should be taken to damage the root-system as little as possible. The thick, fleshy roots of lilies are important, and are always present (on healthy bulbs), never withering completely at the end of the season as happens with tulips and crocuses. Clearly then, when buying new bulbs, it is worth trying to obtain them freshly lifted with plenty of sound roots. The earlier in autumn they can be planted, the more time they will have to produce a sound root-system before winter arrives. They will then be much more likely to establish well. Lilies may also be transplanted early in spring, but more care is needed to check their growth as little as possible.

Surrounding lily bulbs with sharp sand when planting them is a practice that is often advocated. I am dubious of its benefits, and if the soil has been generally well prepared it is almost certainly unnecessary. It may be advantageous where slugs are a problem, but otherwise a mixture of soil, grit and leaf-mould, with a sprinkling of bone-meal, is probably preferable as a planting medium.

The use of fertilizers on lilies has usually been indulged in only sparingly. The Chinese, however, who have grown some of their native lily species for many centuries, have always considered chicken manure to be very beneficial. It may be that its high phosphorus content and tendency to acidify the soil are especially appreciated by lilies, and it should be worthwhile to experiment with its use. Liquid feeding was also used in China, and has indeed been found to encourage strong growth. It would certainly seem that in the past many growers (in Britain, at least) have been too reluctant to feed their lilies, and that moderate amounts of fertilizers may be used to good effect.

(*a*) *Easy lilies for borders and shrub borders*

There are a few Chinese lilies which are reasonably likely to succeed under any normal garden conditions. Probably the most reliable of these are *Lilium regale* and *Lilium lancifolium*. The latter has gained a bad reputation for carrying virus disease, of which it itself shows no obvious symptoms but which may be transmitted by aphids to other lilies nearby. This is not a real problem if it is grown in isolation. In a limy soil, *Lilium henryi* will often flourish. All these species associate well with shrubs, and might be planted towards the front of a shrub border. The tall, weak stems of *Lilium henryi* appreciate the support that branches can give them. The other two lilies will often do just as well in the herbaceous border, as long as the situation is not too open. No lily enjoys exposure to cold winds, and *Lilium regale* will often suffer if its emerging shoots are exposed to late spring frosts.

Lilium davidii, *Lilium dauricum* and *Lilium leichtlinii* var. *maximowiczii* are other moderately easy plants for placing in gaps among shrubs. The form of *L. tsingtauense* now usually encountered in gardens is also quite robust, and is a good plant for light shade in the woodland garden, but is apparently not too demanding and will grow in other situations where there is shelter and partial shade.

Other lilies of section Regalia, such as *LL. sargentiae* and *leucanthum*, may grow quite robustly among shrubs if conditions are to their liking. These are not such easy lilies to grow, however, and great care must be taken to give them a suitable situation. They should be propagated regularly to ensure that stocks are maintained. *Lilium sargentiae* only sets seed in Britain in a hot summer, but compensates for this by producing stem bulbils. Seed should nevertheless still be sown whenever available, to increase the chances of propagating virus-free plants.

(*b*) *Woodland lilies*

The only Chinese lilies which prefer real woodland conditions are the species of *Cardiocrinum*. These normally grow in the shade of trees, and do not enjoy too much exposure to sunlight. They like a soil which is very open and rich in leaf-mould, and resent too much competition from other herbaceous plants. The Chinese species in cultivation, *Cardiocrinum giganteum*, is a majestic plant, taller than any other lily, with beautiful white trumpet-shaped flowers. It is not a lily for small gardens, however, and also has the unfortunate habit of dying after flowering, leaving seeds and a number of small offset bulbs to continue its race. These may take as long as ten years to reach flowering size, so most gardeners will probably continue to see this fine plant only when visiting botanic and other large gardens.

LL. tsingtauense and *distichum* enjoy the kind of conditions which are found at the edges of woodland. Open soil with leaf-mould is undoubtedly their preference, and they do not like too much exposure to direct sunlight. They should not, however, be sited in deep shade under a complete canopy of branches, and like plenty of moisture during the growing season.

(*c*) *Lilies for the peat garden*

The peat garden is undoubtedly the best site for the majority of Chinese lily species. All the species of *Nomocharis* and *Notholirion*, as well as the alpine species of *Lilium*, are usually most easily provided with suitable conditions in a well-drained peat bed. Those who wish to gain inspiration should visit the Royal Botanic Garden, Edinburgh, to see the range of lilies successfully cultivated in the peat beds there.

When constructing a peat garden with lilies in mind, drainage must (as always!) be considered. In fact, the Sino-Himalayan alpine and sub-alpine flora, which is the mainstay of the peat garden, generally prefers free-draining soil. But it is the lilies that are among the most sensitive of plants in this respect. The problem in Britain is that, if it drains sufficiently well not to become too wet in winter, the soil of a peat bed will probably tend to desiccate in summer. Watering is often the only answer to this problem, though careful siting of the peat garden in a semi-shaded area facing north or west, and sheltered from drying winds, will help considerably. Placing the lilies among dwarf rhododendrons or similar evergreen peat-loving shrubs is also beneficial, as their roots will take up appreciable amounts of water during mild periods in winter. It is such damp, mild spells that are the biggest problem, which is why most of the alpine lily species grow better in areas with cold winters. Freezing conditions keep the bulbs dry.

The soil mixture for a peat garden in which lilies are to be grown may have grit as well as peat included in it. Much depends on the qualities of the original soil, but roughly equal quantities of soil, peat and grit will usually produce a satisfactory texture. Such a mixture will be found excellent for Himalayan gentians, *Cyananthus*, *Primulas* and so on, which may be planted with the lilies among dwarf shrubs to make an attractive and mutually supportive community of plants. The smallest lilies, such as *Lilium nanum*, should of course be placed among the lowest shrubs towards the front of the garden, while the taller species should be associated with less dwarf kinds. Particular care should be taken in positioning lilies such as *L. duchartrei* whose stems creep extensively below the soil surface, as they must be given room to wander.

Apart from the *Notholirion* and *Nomocharis* species, the following Chinese lilies are suitable for the peat garden:

L. nanum	*L. henrici*
L. duchartrei	*L. nepalense*
L. papilliferum	*L. taliense*
L. wardii	

L. bakerianum might perhaps also have the best hope of success outside the glasshouse in the peat garden, but it is now too rare in cultivation for many risks to be taken with it. Other species of section Lophophorum would also be peat garden plants, if they ever reach western gardens again.

(*d*) *Lilies for the rock garden*

All the smaller north Chinese lilies are good plants for the rock garden. *Lilium pumilum* is the most commonly available and the easiest to grow, but tends to be rather short-lived. It usually produces plenty of seeds, however, and may be kept going through regular sowings. It enjoys full sun and the sharpest of drainage, but needs plenty of water during its season of growth. Similar conditions will suit *Lilium concolor*, though it is another species which rarely lives very long. The smaller forms of *Lilium dauricum* are also at home in the rock garden environment. The dwarf high-altitude forms of *Lilium formosanum*, usually given the varietal name *pricei*, are very beautiful, and will often flourish in the alpine garden, but need a more moisture-retentive soil than the preceding species.

Most of the other Chinese lily species are rather tall for all but the biggest of rock gardens. *Lilium callosum* and *Lilium cernuum* usually reach about 60cm in height, and so might be included in a garden of moderate proportions.

(*e*) *Lilies under glass*

A few Chinese lilies are not fully hardy in cool temperate climates, and need the protection afforded by a glasshouse. They include *Lilium longiflorum* and *Lilium sulphureum*. *Lilium bakerianum* is probably also more easily grown under glass. A number of other species, such as *Lilium speciosum*, may need some protection in colder climates, and are in any case extremely decorative. They should for preference be planted out in greenhouse borders, but they can also be very satisfactorily grown in large containers, though more effort is then involved. Lilies whose stems creep below ground, as do those of *Lilium nepalense*, must be given adequate space, but have often succeeded in large pots or tubs. Stem-rooting species must be planted sufficiently deeply.

Enthusiasts may also wish to grow some of the smaller lilies in their alpine-houses, where they can be given more care and be more easily seen and appreciated at staging-level. There should be no major problems with this, provided the particular requirements of each species regarding soil composition, drainage and watering are attended to carefully.

(iv) Lily pests and diseases

Lilies are unfortunately prone to attack by a considerable number of pests and diseases. Many of these, however, as is often the case with cultivated plants, are rarely a problem if growing conditions are good and the plants are tended carefully. A few may strike under any conditions, and it is these which constitute the greatest menace. Precautions must always be taken to try to prevent their occurrence, and immediate remedies applied whenever they are seen.

Undoubtedly the worst disorders to which lilies are vulnerable are a number of virus diseases. Some species are more susceptible to these than others, and symptoms vary greatly in severity. *Lilium formosanum* is among the worst affected, with *Lilium speciosum* also easily succumbing. It is, in fact, really preferable for infected lilies to die rapidly, as the infection is then easily identified and diseased plants may be dug up and burned before the virus has much time to spread. But often plants are merely weakened, and may dwindle very slowly, or even show no obvious symptoms at all. If plants appear stunted, with twisted and distorted foliage, which is usually also streaked with yellow, then virus is probably the cause. There is no cure for virus diseases, and as soon as infected plants are noticed they should be immediately removed. The viruses involved are normally spread by aphids, so obviously control of these pests should be as complete as possible. Keeping groups of lilies well separated from each other also helps to slow down the spread of virus disease. The infection is not known to be communicated to seeds; frequent propagation by seed-sowing is therefore of great assistance in maintaining virus-free stocks.

There are also a few common fungus diseases which may affect lilies. They are, however, often a sign of unsatisfactory growing conditions, and should not occur often if sufficient care is taken. If spring weather is particularly wet, *Botrytis* is most likely to be a problem, as it is encouraged by moist conditions (it can be very serious under glass). Usually symptoms appear first on the lower leaves, which develop dark, rounded blotches. The tissue of these infected spots dies, and the disease will then spread up the stem. It may kill all the leaves, and destroy the flower-buds and even rot the stem. It is uncommon for it to infect the bulb, however, and timely spraying with a copper-based or systemic fungicide should give good control. Repeated sprayings can eradicate it entirely. Clearing away all dead stems and leaves at the end of the season will help to prevent the build-up of overwintering spores.

Other fungus diseases attack the bulb in the soil, normally causing it to rot completely. They should not be a problem if soil conditions are good, but are often introduced when new stock is bought in. Any newly-acquired bulbs that show any signs of damage or disease should therefore be treated before planting, by removing affected scales and soaking in a weak solution of systemic fungicide. Once the fungi have entered the soil, and are causing the deaths of lily bulbs, the most effective cure is to dig up all the lilies in the infected area and discard any that are showing signs of infection. The rest may then be treated with a fungicide and replanted elsewhere. No lilies should be planted in the affected soil for a minimum of three years.

These diseases are the most serious troubles of lilies, but certain pests may also cause great damage. Aphids have already been mentioned, and apart from spreading virus infections may cause quite severe damage through their own juice-sucking activities. Whenever they are seen on lilies, they should be removed. Anything more than a light infestation will require the use of some kind of spray. There are many different species of aphid that may attack lilies, but they are all easily killed by any of the commonly used insecticides. Repeat sprayings will normally be necessary. Systemic insecticides are reported to give very good protection.

Another insect, the Lily Beetle, is a serious pest in certain restricted areas. I had trouble with it in a garden at Woking (not very far from Wisley). The adults are very conspicuous, as they are bright scarlet and quite shiny, with black legs. The larvae are not so obvious, as they are a muddy yellow colour and cover themselves with their own excrement, but the damage they do will quickly be noticed. Both they and the adults feed voraciously on all the green aerial parts of lilies. A small infestation may be picked off by hand, but an insecticide spray should be used if more than two or three beetles are seen.

Other insects and their larvae may cause occasional damage to lilies, but are rarely serious primary pests. It is more common for them to take advantage of already-existing damage. Wireworms and leatherjackets may attack lilies in some gardens, and can be controled with BHC dust worked into the soil.

Slugs and snails are another story. After aphids, as carriers of viruses, they are the deadliest enemies of the lily. They may eat any portion of the plant, above ground or below, and are quite capable of causing complete destruction. Seedlings and young plants are particularly vulnerable. The commonly used slug baits and pellets provide little or no protection, as the creatures may well find the lily before they eat the poison. They are, moreover, highly toxic to the gardener's allies, birds and hedgehogs, and can therefore do much more harm than good. One hedgehog is worth whole packets of slug pellets, and will certainly destroy more slugs. There is now a product available called 'Fertosan' slug killer, which is a soluble white powder toxic only to molluscs. It is a contact killer, not needing to be eaten by its victims. It may be scattered as dry powder or dissolved and watered onto the soil. It seems to have longer-lasting effect if watered on. It kills slugs and snails almost instantly, and if applied regularly, every two or three months, will give continuous, more or less total control. Its only drawback is that it can scorch tender shoots, and should not be applied directly to seedlings, especially as powder.

There are a number of other pests and diseases which may be encountered when growing lilies, but they are generally neither common nor severe. If healthy bulbs are obtained at the start, planting sites are thoroughly prepared, and attention is paid to looking after the lilies and clearing away dead stems at the end of the season, then few problems should be encountered. Regular sowing of seeds will enable healthy new plants to be always available even if old ones succumb to virus. Although they cannot in general be considered 'easy' plants, most Chinese lilies are by no means impossible to cultivate successfully.

(v) Reference guide to Chinese lilies in the garden

SPECIES	HEIGHT	BORDER	WOODLAND	ROCK GARDEN	PEAT GARDEN	GLASSHOUSE
Lilium:						
L. concolor	Small			*		
L. dauricum	Medium	*E		*E		
L. brownii	Tall	*				
L. regale	Medium	*E				
L. formosanum	Small – Tall	*		*		*
L. longiflorum	Medium	*				*
L. leucanthum	Tall	*				
L. sulphureum	Medium	*				*
L. sargentiae	Tall	*				
L. bakerianum	Medium				*	*
L. nanum	Small				*	
L. brevistylum	Small				?	

SPECIES	HEIGHT	BORDER	WOODLAND	ROCK GARDEN	PEAT GARDEN	GLASSHOUSE
L. lophophorum	Small				?	
L. souliei	Small				?	
L. sempervivoideum	Small				?	
ssp. *amoenum*	Small				?	
L. paradoxum	Small				?	
L. medogense	Small				?	
L. henrici	Medium				*	*
L. speciosum	Medium	*				*
L. henryi	Tall	*E				
L. rosthornii	Tall	?				
L. nepalense	Medium				*	*
L. wardii	Medium				*	
L. stewartianum	Small				?	
L. habaense	Small				?	
L. taliense	Tall				*	
L. duchartrei	Medium				*	
L. davidii	Medium – Tall	*E			*	
L. leichtlinii	Tall	*E				
L. lancifolium	Medium	*E				
L. pumilum	Small	*E		*E		
L. cernuum	Medium	*	*	*		
L. callosum	Small			*		
L. papilliferum	Medium				*	
L. fargesii	Small				?	
L. xanthellum	Small				?	
L. tsingtauense	Medium	*	*			

SPECIES	HEIGHT	BORDER	WOODLAND	ROCK GARDEN	PEAT GARDEN	GLASSHOUSE
L. distichum	Medium		*			
L. martagon	Medium	*E				
Cardiocrinum:						
C. giganteum	Tall		*			
C. cathayanum	Medium – Tall		?			
Nomocharis:						
N. saluenensis	Medium				*	
N. aperta (inc. N. forrestii & N. biluoensis)	Medium				*	
N. pardanthina	Medium				*	
N. farreri	Medium				*	
N. meleagrina	Medium				*	
N. basilissa	Medium				?	
Notholirion:						
N. bulbuliferum	Medium – Tall				*	
N. campanulatum	Medium – Tall				*	
N. macrophyllum	Small				*	

KEY:
E: easily-grown lilies
? indicates the probable site for species not in cultivation

Small: up to about 60cm (2ft) in height
Medium: up to about 1.6m (5ft) in height
Tall: up to 2m (6ft) or more in height

CHAPTER THREE

The history of Lilies in China

It has long been known in the west that lilies have been cultivated in China since ancient times, yet no detailed and accurate account of the history of lilies in China has until now appeared in a western language. Although E. H. Wilson, for example, devoted a chapter in *The Lilies of Eastern Asia* to 'History', only one paragraph referred to their history in China and Japan, and even this was not entirely accurate. Subsequent writers have made no better attempts. This seems a pity, for the history of garden plants is a fascinating subject. It can add greatly to our appreciation of the plants we grow to know something of their antecedents.

To the Chinese, lilies are not only beautiful, but have their practical uses as well. The bulbs of several species are esteemed as culinary delicacies, and are also considered to have medicinal properties – they are said to be cooling, moistening and calming in effect, and are prescribed for such complaints as chronic coughing, certain disorders of the blood, neurosis and sleeplessness. In fact, it is probable that lilies were originally valued by the Chinese for their efficacy as medicine, and were cultivated as much for the virtues of their bulbs as for the beauty of their flowers. The earliest certain references to lilies in Chinese literature occur in herbals or pharmacopoeias which describe the uses of medicinal plants.

The oldest text of this kind now extant is the *Shen Nong Bencao Jing* or *Divine Husbandman's Classic of Herbal Medicines*. Though it is attributed to the legendary deity of its title, who was credited with originating agriculture and ascertaining the uses of all plants, the real authorship of this book is unknown, and its precise dating is uncertain. It was compiled not later than the second century AD, but was probably based on earlier writings and traditions, and must include a great deal of very much older material. In about AD 520 it was edited by Tao Hongjing, who also wrote a supplementary treatise, the *Ming Yi Bie Lu (Additional Records of Famous Doctors)*. After this it formed the basis upon which virtually all Chinese *materia medica* were compiled, and was extensively quoted in many major works. This ensured its survival, for it was later lost as an independent text, and has been reassembled from quotations. It lists 365 medicines, of which 252 are herbal. Among these is *Bai He*, the Chinese name for certain kinds of lily. It is therefore likely that lilies have been used medicinally by the Chinese for at least two millennia.

There is no description of the physical appearance of *Bai He* in the *Shen Nong Bencao Jing*. The name, however, may be translated 'Hundred United', and obviously refers to the appearance of the bulb. Tao Hongjing wrote: 'The root is like garlic, with several tens of segments joined together. People steam or boil and eat them. It is said that at first they are earthworms which knot themselves together and are transformed.' He also says that 'it grows in the valleys of streams in Jing Zhou', an administrative region covering a large part of central China and corresponding approximately to the two modern provinces of Hunan and Hubei. The story of the earthworms is an interesting example of Chinese ideas regarding the transformations of organisms, and is not without parallel in pre-modern European thought.

The earliest description of the above-ground portion of the plant occurs in the *Xin Xiu Bencao (Newly-revised Pharmacopoeia)* which was compiled at Imperial command by Su Gong and others, and was presented to the Emperor Gao Zong of the Tang dynasty in AD 659. This is claimed to be the first government-sponsored pharmacopoeia ever produced. Of *Bai He* it says: 'There are two kinds of this medicinal plant. One kind has narrow leaves and red flowers. The other has large leaves, a tall stalk, a large root and white flowers, and is good for medicinal use.' The *Shi Liao Bencao (Pharmacopoeia of Edible and Curative Herbs)* of the early eighth century says 'those with red flowers are called *Shan Dan* and are not suitable for eating'. The *Ri Hua Zi Bencao (Pharmacopoeia of Master Ri Hua)* of about AD 970 also distinguishes between red and white lilies, giving different uses for the two kinds.

A more detailed description appears in the *Bencao Tu Jing* or *Illustrated Pharmacopoeia*, which dates from about 1080 (its original illustrations have unfortunately been lost). This says that:

> *Bai He* . . . produces shoots in spring which grow to several feet in height. The stalk is as thick as *Arundinaria* bamboo, bearing leaves all round which are like chicken spurs, and also similar to willow leaves. They are green in colour, slightly purple near the stem. The extremity of the stem is bluish-white. In the fourth and fifth months [of the Chinese lunar calendar, approximately equivalent to June and July], it opens its red and white flowers, which are like those of Pomegranate but with a wider mouth. The root is like garlic, with several layers, and composed of twenty or thirty segments. The roots are collected in the second and eighth months, and dried in the sun. People also steam and eat them. They are beneficial to the *qi* [vital energy]. There is also a kind which has orange flowers with black spots, narrow leaves and black seeds [i.e. bulbils] among the leaves. It is not suitable for medicinal use.

There is another interesting description in the *Bencao Yan Yi (Herbal*

Medicines Explained), completed in 1116 by Kou Zongshi, who was at one time an official in what is now northern Hunan province. His account is as follows:

> The stem is about three feet tall. The leaves are like large willow leaves, crowded all around the stem and upwardly inclined. At the top of the stem open pale yellowish-white flowers, which incline downwards to each side and bear long stamens; the centre of the flower is tinged with the colour of sandalwood. Each stem carries some five or six flowers. The seeds are purple, and round like those of the Chinese parasol tree [*Firmiana simplex*]. They grow between the stem and the leaves, one seed to each leaf, and not in the flowers, which is most strange. The root is the part used as *Bai He*. It is white, and shaped like a pine cone, [with scales] crowded all around. The shoot is produced from its centre.

Thus it can be seen that by the twelfth century several different kinds of *Bai He* were recognized by the Chinese, some with white and some with red flowers, some with stem bulbils and some without, and differing in other characteristics. It would appear from these descriptions that the most commonly used lily was tall and had white flowers, and that lilies with red or orange flowers were not held in very high esteem, some writers even stating that they were not fit for eating. Undoubtedly the usual source of *Bai He* bulbs for medicinal purposes throughout this early period was one of the white trumpet lilies of section Regalia, of which the most common and widespread is *Lilium brownii*. The descriptions seem to accord well overall with the appearance of this lily, as does the area given for its distribution. It is therefore generally accepted that it is this species that is the white-flowered *Bai He* of the early Chinese pharmacopoeias. The description in the *Bencao Yan Yi*, however, says that 'seeds' grow in the leaf axils, which is not normally true of *Lilium brownii*. Either this account is inaccurate, or one of the bulbiliferous trumpet lilies is intended. Possibly Kou Zongshi may have been acquainted with *Lilium sulphureum*, which has stem bulbils and flowers of the coloration described. It is possible that plants of this species could have been brought down the Yangtze River from Sichuan to the area of Hunan where Kou was an official, for this was on the banks of a major tributary of the Yangtze amid the network of waterways around Lake Dongting.

The bulbiliferous lily with 'orange flowers with black spots' is quite obviously the tiger lily, *Lilium lancifolium*. This later became the most commonly cultivated and eaten lily in China, and is now the main source of the bulbs used as medicine, so that it is odd that the *Bencao Tu Jing* states that it 'is not suitable for medicinal use.' This assertion was repeated, often through quotation of this source, in most subsequent Chinese pharmacopoeias at least until the second half of the eighteenth century. As it remained unchallenged for so long, it seems most unlikely that *Lilium lancifolium* could have been widely grown for its bulbs in China until comparatively recently. The suggestion, frequently repeated in western writings (even by such authorities as E. H. Wilson and P. M. Synge), that it may have been in cultivation for longer than any other lily, must therefore be incorrect. Probably westerners visiting China in the nineteenth and twentieth centuries, and seeing this lily being grown there as *Bai He*, thought that this was the plant which had always been used and grown under that name. It is quite clear from detailed examination of the early sources that this was not the case. It is not only the written descriptions which confirm this. In 1249 a new edition of the official pharmacopoeia of the Song dynasty was printed, complete with illustrations. This was the *Chongxiu Zheng He Jingshi Zhenglei Beiyong Bencao (The Authorized, Classified and Practical Pharmacopoeia of the Zheng He Reign-period: Revised)*. Fortunately, at least one copy of the original edition of this book has survived. Its block-printed illustrations of *Bai He* clearly depict a trumpet lily, and not *Lilium lancifolium*. They could well be intended to portray *Lilium brownii*, and it may safely be assumed that this species was the usual source of lily bulbs for medicinal use at that date.

The early texts previously quoted also refer to a lily with red flowers and narrow leaves, which the *Shiliao Bencao* says is called *Shan Dan*. This seems to have been distinct from the tiger lily, and indeed later usage applies the name *Shan Dan* to either or both of *Lilium pumilum* and *Lilium concolor*. There is no reason to think that this was not so during these early times also; both these species have the narrow leaves and red flowers of the description, and grow wild in large areas of China.

From the medicinal texts it is impossible to be sure of whether the lily bulbs used medicinally were normally collected from the wild or cultivated. In early times they must usually have been dug from the wild, for mention of the cultivation of lilies is difficult to find in texts of early date. The earliest undisputed reference seems to be that found in the agricultural treatise *Si Shi Zuan Yao*, or *Compendium of Essential Tasks for the Four Seasons*, which was written at some time during the period between about AD 900 to 960. As its name suggests, it sets out in chronological order agricultural activities for the year. Under the heading 'Second Month' (which would be approximately equivalent to April), the following entries appear:

> **To plant *Bai He*:** This crop benefits greatly from chicken manure. Make each trench five *cun* [Chinese inches] deep, put in chicken manure, and put *Bai He* scales on top of the manure, in the same way as when planting garlic. *Bai He* is transformed from worms, yet on the contrary it likes chicken manure. The reason for this cannot be understood.
>
> ***Bai He* Flour:** Take the roots and dry them in the sun. Pound them into flour and sieve fine. This is very beneficial to people.

It must be borne in mind that the use of chicken manure, as advocated here, was for growing lilies purely for their bulbs. While liberal use of manure may encourage the bulbs to grow to a large size quickly, it is generally accepted that the application of fertilizers to lilies grown for ornamental purposes should be indulged in only in moderation. Overfed bulbs may be more susceptible to diseases, and do not necessarily produce better displays of flower. E. H. Wilson criticized the demand for extra-large bulbs current among western enthusiasts in his time, which had led Japanese growers for the export market to feed their lilies richly so as to produce bulbs of maximum size. With his considerable experience of lily bulbs collected from the wild, he wrote that 'a firm, solid bulb of moderate size will be found more healthy and will give results more satisfactory than a large, loose and flabby bulb.' Having given this warning, it might still be worth trying the use of chicken manure on lilies when the application of fertilizer is thought desirable.

It is also interesting to note from the passages quoted above, that the Chinese were already familiar at this early date with the method of propagating lilies by separating the bulb scales. As it is this method that the *Si Shi Zuan Yao* advocates, it is probable that the lily its author had in mind was the non-bulbiliferous *Lilium brownii*, or at least not *Lilium lancifolium*, which is more easily propagated by means of its stem bulbils.

The Chinese inch or *cun* was normally somewhat longer than the Imperial inch (though its exact length varied in the past not only from time to time but also from place to place!). The depth of trench advocated is therefore quite considerable for separated scales, though it must be noted that they were to be put in on top of the layer of manure.

Making flour of lily bulbs is possible because they have a high starch content. The idea of eating the bulbs seems strange to western minds, but the Chinese are very adventurous in their eating habits. It is doubtful, however, whether many present-day lily enthusiasts are likely to have such a surfeit of bulbs that they may feel tempted to add them to their menu!

It is certain, then, that lilies began to be cultivated in fields for their edible bulbs at least as early as the tenth century. It may also be that they were already being grown for ornamental purposes in China during this early period. There is no doubt that they were appreciated for more than just the nutritional qualities of their bulbs: the beauty of their flowers had not gone unnoticed. A poem in praise of *Bai He* has survived down to our own times from as early as the mid-sixth century. Composed by the Emperor Xuan of the short-lived Later Liang dynasty, it extols the purity of colour of their flowers:

> Their leaves cluster layer upon layer;
> Their flowers open immaculate,
> Cupping the dew or downward inclined.
> They sway with the motion of the air.

The great Tang dynasty poet Wang Wei (AD 701–760) also composed lines on *Bai He*. But it is not clear from these poems whether the lilies written about were cultivated or wild. It must be considered likely, however, that plants which were obviously appreciated for their fine flowers, and were cultivated in fields, would also have been grown in gardens. There is evidence of this in a verse by Su Dongpo (1036–1101), who wrote the following couplet:

In front of the hall are planted *Shan Dan* –
A pattern of colour like a cornelian tray.

This seems also to be the earliest certain reference to the cultivation of *Shan Dan* (rather than the white-flowered *Bai He*). As the small bulbs of the red lily were not rated so highly for medicinal usage, it was probably most commonly cultivated for ornament. It may well be that it was more popular in gardens than its white-flowered relative, as it seems to be mentioned more frequently in poetry of the eleventh to thirteenth centuries.

So by the middle of the thirteenth century at least three different lilies were distinguished in China. The first of these, and the most important one for medicinal usage, was a tall white-flowered species called *Bai He*, which was cultivated at least from the early tenth century. This was *Lilium brownii* (though some of the other white trumpet lilies may perhaps have been confused with this species and included with it). *Lilium lancifolium* was also known, though no separate name for it is recorded, nor is there any clear indication that it was cultivated during this period. Finally, there was the small, red-flowered lily called *Shan Dan*. It is impossible to tell from the brief descriptions exactly which lily this was, and it is indeed very likely that the name was used for more than one species. The common and widespread *Lilium concolor* and *Lilium pumilum* would probably have been the lilies to which this name was most frequently applied, but *Lilium callosum* and perhaps other species with orange-red flowers may also have been included here. *Shan Dan* were grown in gardens at least as early as the eleventh century.

Chinese horticulture must have suffered a severe setback during the twelfth and thirteenth centuries, when there was protracted warfare. This began with the invasion of Song dynasty China by the Jürched from Manchuria, which resulted in the loss to the invaders by 1126 of virtually all of China north of the Yangtze, with fighting continuing sporadically until the 1160s. Half a century later came even greater turbulence, as China felt the effects of the expansion of Mongol power. The armies of Genghis Khan began to harass the Jürched kingdom of the Jin dynasty as early as 1211, and by 1234 his son Ögödei had completely taken over control of the northern half of China. The Southern Song dynasty was able to maintain its hold on China south of the Yangtze for another 40 years, but finally its increasingly precarious existence was brought to an end by the forces of Khubilai Khan early in 1279. The destruction caused by this prolonged fighting was enormous, especially in north China, where the Mongols massacred many thousands of people and turned large areas of cultivated land into pasturage for their horses and herds. It was not until the decline of Mongol power had led to their overthrow in China and to the re-establishment of a native Chinese dynasty, the Ming, in 1368, that there seems to have been a renaissance of the peaceful arts of plant cultivation. Even so, it was only towards the end of the period of Ming rule, in the late sixteenth and early seventeenth centuries, that any major new contributions were made to the Chinese literature on plants and their uses.

It was perhaps a sign of the troubled times that the most significant work relating to plants to be written during the early Ming dynasty was the *Jiu Huang Bencao* or *Famine Herbal*, compiled by a prince of the Imperial family and printed in 1406. It describes more than 400 wild plants which could be eaten in times of famine. It is recorded in a preface to the book written by the prince's secretary that all the plants were grown and tested before the book was written. There is a passage on *Bai He* in the text, though much of it is derived from the early pharmacopoeias, and there is little in the way of new material. The red-flowered *Shan Dan* is briefly mentioned, but is said not to be suitable for use. As the plants included in this work were to be gathered from the wild when crops had failed, it must be presumed that the prince knew *Bai He* as a wild plant rather than as a crop, which may mean that it had gone out of cultivation in his time. But a commentary added in about 1630 by Xu Guangqi says: 'I have tasted this. The root is an excellent vegetable, and need not be used only in time of famine.' It was certainly grown in fields in about 1500, for an agricultural treatise of that time, the *Bian Min Tu Zuan (Illustrated Compendium to Benefit the People)*, contains an entry relating to the planting of *Bai He*.

Whatever the situation was in the early 1400s, it is likely that by the latter half of the sixteenth century many of the lily bulbs used medicinally came from cultivated plants. This can be deduced from the fact that in the *Bencao Gang Mu (Materia Medica Catalogued)*, the entry for *Bai He* appears for the first time in the 'Vegetables' section, and not among the wild herbs. The greatest of all the old Chinese herbals, the *Bencao Gang Mu* represented virtually a lifetime of labour by its author, Li Shizhen, and was completed in 1578 (though not printed until 1590). It contains a great wealth of material, both derived from earlier pharmacopoeias and added from Li Shizhen's own experience: he had tested virtually all of the almost 2000 medicines described in his book. He distinguishes clearly between three different kinds of lily:

> Those with short, broad leaves rather like bamboo leaves and white, nodding flowers, are *Bai He*. Those with long, narrow leaves which are pointed like willow leaves, and red flowers which do not nod, are *Shan Dan*. Those which have leaves and stems like *Shan Dan* but are taller, with nodding, red flowers which are tinged with yellow and spotted with black, and which first set seeds in the leaf axils, are *Juan Dan*. *Juan Dan* produce seeds in the fourth month but flower in autumn, and have roots like *Bai He*, while *Shan Dan* flower in the fourth month and have small

> roots with fewer scales. Thus there are three kinds of this one type [of medicinal herb].

These descriptions clearly indicate that Li Shizhen knew *Lilium brownii* as *Bai He*, *Lilium concolor* as *Shan Dan*, and *Lilium lancifolium* as *Juan Dan*. The statements that the flowers of *Shan Dan* 'do not nod' and that it has 'long, narrow leaves', make it extremely unlikely that any lily other than *Lilium concolor* could be intended. This is the earliest description of *Shan Dan* which is detailed enough to permit such precise identification. It is likely that there was confusion at this time about the identity of the various kinds of *Bai He*, which made it necessary for Li Shizhen to stress the distinctions between the three that he recognized. It is interesting that, despite their superficial dissimilarities and the fact that their medicinal efficacies were believed to be very different, the various *Lilium* species were obviously considered to be very closely related. Li Shizhen did, however, accord a separate entry to *Shan Dan*, immediately following that for *Bai He*. He noted that it was not considered to be as useful medicinally, and added that: 'The people of Yan and Qi [Hebei and Shandong] collect its unopened flower buds, dry them and sell them, calling them "Red Flower Vegetable"'. I am unable to say whether this still happens today, but it is interesting that the flower buds of several species of the closely related genus *Hemerocallis* are currently used as a vegetable in this manner, and are called 'Yellow Flower Vegetable'. I have personally eaten them in China on more than one occasion.

Li Shizhen's approach to the study of medicinal plants was more rational than had often been the case with earlier writers. This is shown clearly by the care which he takes in describing the distinguishing features of the three lily species, and is also demonstrated by his scepticism about the idea that lily bulbs derived from worms:

> Those which occur wild in the hills grow every year from their persistent roots. They cannot all be transformed from worms, for worms occur in many places, and I have not heard that there are *Bai He* in all those places. I fear that this saying is just wild speculation.

It is, however, very probable that he ought to have been aware of more than just the three lilies that he described. Only 40 years or so after he completed his *Bencao Gang Mu*, there appeared another major work, which listed a much larger number of kinds. The *Qun Fang Pu*, or *Register of All Sweet Flowers* is a vast compilation of information on plants. As its title indicates, it is largely concerned with their ornamental value, so it may be assumed that all the many plants it lists were grown in Chinese gardens at the time when it was written (the early seventeenth century). It is divided into various categories, headed 'Flowers', 'Trees' and so on, and it is revealing that *Bai He* is entered under 'Fruit'! This must be taken to indicate that *Lilium brownii* was still chiefly cultivated at this time for its edible bulb, rather than its beauty as a flowering plant.

The entry for *Shan Dan*, however, does appear in the section headed 'Flowers', and says:

> Its root is similar to that of *Bai He*, but is smaller in size, with fewer scales. It can be eaten. . . . It flowers in the fourth month. There are two kinds, red and white. . . . There is also one which flowers throughout the year, called 'Four Seasons *Shan Dan*'. It sets small seeds. The people of Yan and Qi collect its flowers and dry them in the sun, and call them 'Red Flower Vegetable'. . . . Another kind is four or five feet tall with flowers like *Hemerocallis fulva*, which are as big as saucers and are red with black spots. The petals are revolute. Each leaf bears a seed [bulbil], and it is called 'The Flower which Turns its Head to See its Offspring', and also called the 'Foreign *Shan Dan*'. Another kind is a foot or so tall with flowers of the colour of cinnabar. Strong specimens carry two or three scentless flowers on one stem.

This is followed by another entry, for a plant called *Wo Dan*, which is clearly another kind of lily:

> It is also called *Shan Dan*. . . . Its flowers are smaller than those of *Bai He*. . . . When it opens its flowers, they are of the reddest colour.

This is important information about the kinds of lilies that were cultivated in China during this period, though unfortunately the descriptions given are not detailed enough to make identification easy. It is obvious that the lily 'with flowers like *Hemerocallis fulva*' and 'seeds' among the leaves is *Lilium lancifolium*. 'The Flower which Turns its Head to See its Offspring' is a fanciful but accurate depiction of the Tiger Lily, with its nodding flowers apparently looking downwards towards the bulbils in its upper leaf axils. But even this description is not entirely free of problems. Why should a lily species which is native to China and had been known and described long before this period now have the name 'Foreign *Shan Dan*' applied to it? In fact, slightly later writings, which will be considered below, do not seem to consider that the name belongs to this lily. I think it should probably correctly be attached to the very similar *Lilium leichtlinii* var. *maximowiczii*, which could easily have been confused with the Tiger Lily, the most obvious difference being that it lacks stem bulbils. Although *Lilium leichtlinii* var. *maximowiczii* is native to parts of north-eastern China, it does not seem to be common, except perhaps in some mountainous areas of Manchuria. But Manchuria did not form part of the Ming empire, so that it would not be surprising if it first came into cultivation in Chinese gardens from Japan. This would explain the epithet 'Foreign' being given to it.

The plant called *Wo Dan*, to judge from later usage of the name and from the description of the colour of its flowers, was probably *Lilium concolor*. It is here clearly stated that *Shan Dan* was an alternative name for this lily, so that it is now made obvious that the name *Shan Dan* was applied to more than one species.

But if *Lilium concolor* in this work is given the name of *Wo Dan*, this poses the problem of which lily the name *Shan Dan* was intended to cover. There is little in the text to assist in answering this question, except that it is said that there were both red and white colour forms of this kind of lily. From later usage of the name it might be assumed that *Lilium pumilum* was intended, and it is indeed the case that white forms of this lily are known to occur (no white form of *Lilium concolor* is known). They may well have been picked out by Chinese gardeners and propagated as a curiosity.

The 'Four Seasons *Shan Dan*' defies explanation. There is no known lily which behaves in this way, except that in cultivation it is possible to force many lily species to flower out of season. Perhaps such forcing was being carried on in China at this time, and thus gave rise to this name. Possibly the process came into use in order to allow the provision of 'Red Flower Vegetable' throughout the year (though the flowers when dried might be expected to keep for some time). This is, however, highly speculative, and in the absence of further evidence must be left unresolved.

This leaves just one kind of lily still to be identified, that with 'flowers the colour of cinnabar'. Perhaps this was intended to describe the rather dull orange-red coloration of the flowers of *Lilium callosum*. This lily is quite widespread and moderately common in eastern China, with an unusually large range from north to south, and may well have been noticed by Chinese gardeners. Its flowers are scentless, as the description says, and it tends to have rather few flowers to a stem.

Thus by about 1621, when the *Qun Fang Pu* was completed, there were about half a dozen species of lily in cultivation in Chinese gardens. These were all the common lilies of central and east China, *LL. brownii, concolor, lancifolium, pumilum* and probably also *L. callosum*. One described as 'Foreign' was also grown, which must almost certainly have come from Japan and may have been *L. leichtlinii* var. *maximowiczii*. Variant colour forms were already being selected and presumably also maintained in cultivation. Propagation techniques included scaling of the bulbs, and seeds and bulbils would certainly also have been used, as the Chinese clearly understood that new plants could be grown from these.

But for producing large quantities of bulbs the separation of the scales was undoubtedly the normal method used in China. Virus disease must not have been a major problem. The *Nong Zheng Quan Shu (Encyclopaedia of Agricultural Practice)* by Xu Guangqi, published in 1639, gives the following instructions for cultivating *Bai He*:

> They like fertile soil. Add chicken manure, and hoe thoroughly. In spring, take large bulbs, split them up and scatter them in trenches, as for planting garlic, one segment every five *cun*. After two and a half months, weed them completely three times, for if they are not weeded they will not grow. After three years they will be as big as wine-cups. If they are kept well-watered, then they will produce magnificent flowers, whose pure

> fragrance will fill a whole courtyard. They may also be divided at the autumnal equinox.

Additionally, Xu quotes the descriptions of the three kinds of lilies from the *Bencao Gang Mu*, and adds that: 'There is also a kind which is somewhat green in colour, and flowers very late. It is commonly called *Zhen Bai He*.' This must have been another species of section Regalia. Judging by its late period of flowering, it may have been *Lilium sargentiae*, which often has greenish-tinged flowers, but the description is not detailed enough to be certain of this.

The *Bu Nong Shu (Augmented Treatise on Agriculture)* of about 1658 gives further details of the cultivation of lilies as a crop. It recommends planting them between rows of mulberry trees, where they will not affect the growth and care of the trees, and says that they may be harvested every one, two or three years. It also states that they were at that time commonly cultivated in the area north of Hangzhou, in Zhejiang province (near Shanghai), and were considered a speciality of the region. The flowers are described as 'fragrant and immaculate', so it would appear that the species grown was still *Lilium brownii*.

The fall of the Ming dynasty in 1644 and its replacement by the Manchu Qing dynasty did not cause as much disruption as the troubles which had preceded the foundation of the Ming. The early period of Qing rule was a time of peace, prosperity and revitalization. Combined with the fact that many of the scholar-officials who had served the Ming emperors refused to work for the new rulers and went into retirement, this must have encouraged the growth of interest in ornamental gardening. Certainly a number of important works on gardens and their plants were written during the seventeenth and early eighteenth centuries.

The most outstanding of these from a purely horticultural viewpoint is the *Hua Jing* or *Mirror of Flowers*, completed in 1688. This lists garden tasks month by month and explains gardening techniques (including watering, pest control, and many different methods of propagation, such as grafting, layering and air-layering, division and sowing of seed). It then goes on to describe more than 350 kinds of ornamental plants. Among these are two separate groups of lilies.

The first group is listed under the heading *Shan Dan*. The alternative name *Wo Dan* is given, and the description is almost certainly of *Lilium concolor*:

> The flower colour is vermilion-red, such as no other flower can equal. Strong specimens have three or four flowers to a stem, which not only are not fragrant, but moreover fade quickly, so that even the succession of flowers lasts only a few days. Its nature is similar to that of *Bai He*. There are also yellow and white ones, which are renowned as rare varieties. They should be propagated by division in spring, but they also produce small seeds. They flourish if watered and fertilized well – chicken manure is especially efficacious.

The difficulty here is that no white form of *Lilium concolor* is known. Probably this white lily was a form of *Lilium pumilum*: such a lily is known to exist, and as we have seen above, was already cultivated in China. As this variety is said to be rare, it may be that the author, Chen Haozi, had heard of it but never seen it, or that he simply failed to recognize that it was a form of a different species. Yellow colour forms of both *Lilium pumilum* and *Lilium concolor* occur in the wild and in cultivation.

The text goes on to describe 'another kind, called "Foreign *Shan Dan*". Its root and leaves are of the same type as those of *Bai He*, but its flowers are red with black spots. The root tastes bitter, and grows easily, so this kind is not much valued'. As there is no mention here of stem bulbils, and as *Lilium lancifolium* is separately recorded elsewhere (see below), this is further evidence that this 'Foreign *Shan Dan*' was probably *Lilium leichtlinii* var. *maximowiczii*.

The second group of lilies is discussed under the heading *Ye He*, a name which seems to have been used for *Lilium brownii* var. *viridulum* to distinguish it from the type variety. *Bai He* is given as another name for the same plant, and the description accords with this identification. The flowers are stated to be 'honey-coloured', which is usually the case at least with newly-opened flowers of var. *viridulum*. The passage continues with descriptions of several other lilies:

> There is also one named *Tian Xiang* ['Heavenly Fragrance']. Its root is like that of *Shan Dan* but double the size. It has numerous scales tightly wrapped together, just like a white lotus flower. It is sweet, and can be eaten. Another kind is called *She Xiang Hua* ['Musk-Scented Flower'], which is similar to *Tian Xiang*, but differs in being shorter and having more numerous leaves, and in flowering in the fourth month. *Tian Xiang* flowers in the sixth month. Another kind is like *Hemerocallis fulva*. . . . It is commonly called '[The Flower which] Turns its Head to See its Offspring'. . . . Its root is like *Bai He*, but is bitter in flavour and not good to eat. *Bai He* grow up once in a year. The outer scales of large ones may be peeled away, boiled, and eaten, leaving the small inner hearts. If these are replanted in rich soil, then in spring they will send up new shoots just as before. If chicken manure is worked in around their roots they will grow strongly, and they should also be watered regularly with liquid fertiliser.

Although the description is not very full, it is reasonably certain that *Tian Xiang* is the Japanese *Lilium auratum*. The bulbs of this lily are known to have been eaten in Japan, and the evocative description of their appearance fits perfectly. The rather late period of flowering would also be in agreement with this identification (the sixth month would approximate to August). The name was certainly used at a later date for this species. Again judging by later usage and by the flowering period, *She Xiang Hua* ought to be *Lilium longiflorum*, which may also have reached the mainland of China by way of Japan. *Lilium lancifolium* is easily recognizable. It is a great pity that *Lilium brownii* has never taken to cultivation outside China so well as to make it possible to treat it as is recommended at the end of this passage!

Taking into account all the species mentioned for the first time here, there were by this period at least seven or eight different species of *Lilium* in cultivation in Chinese gardens. Most of these were native species, though some had come from Japan. Although lily bulbs were obviously still greatly appreciated as a culinary delicacy, many of these species were grown solely for their flowers. They must undoubtedly have made a considerable contribution to the beauty of the gardens of China.

By 1765 it would appear that lilies were so well established as cultivated plants that it had become normal for bulbs for medicinal use to be of cultivated origin. In that year was printed the *Bencao Gang Mu Shi Yi (Omissions from the 'Materia Medica Catalogued' Made Good)*, in which there is an entry for 'Wild *Bai He*'. This was not, in fact, an omission on the part of Li Shizhen, it was simply that in his time *Bai He* were commonly collected from the wild, whereas by the mid-eighteenth century this was no longer the case. Interestingly, this text also mentions 'Tiger-Skin *Bai He*', which must surely be the Tiger Lily, saying that: 'it kills people who eat it'. So *Lilium lancifolium* would seem still to have been thought inedible. Modern Chinese works on medicinal plants state that this species has been cultivated for more than a hundred years at Yixing in Jiangsu province, so it probably first began to be eaten in China at some time between about 1800 and 1850.

The history of the cultivation in China of lily species other than those already mentioned above is obscure and almost certainly quite short. In fact, the majority of Chinese lilies have probably never been cultivated in their native land. *Lilium davidii*, however, was widely used as *Bai He* in the provinces of Sichuan and Yunnan, and George Forrest found that it was grown as a field crop around Dali, Lijiang and Tengchong in Yunnan when he was collecting there during the early decades of this century. It is still cultivated in various areas in the west of China, as far north as Lanzhou in Gansu province.

The lily once known as *Lilium centifolium*, now reduced to a variety of *Lilium leucanthum*, was also found by a western collector being cultivated in China. It was in 1914 that Reginald Farrer obtained its seeds from the garden of a cottage near Siku (Zhouqu) in southern Gansu. It was probably never widely grown, however.

In recent times, several more native lilies have begun to be cultivated in China, if only in the various botanic gardens that have been established there this century. At the Zhongshan (Sun Yat-sen) Botanic Garden in Nanjing, for example, about a dozen species are grown, including the rather uncommon *Lilium tsingtauense* and the Chinese form of *Lilium speciosum*, the variety *gloriosoides*. There will no doubt continue to be an increase in the number of species cultivated by the Chinese.

The lily now most commonly grown on a large scale in China, and the main source of bulbs for medicinal use, is *Lilium lancifolium*. The main centres of its cultivation are in Jiangsu province, but it is quite commonly grown in other parts of the country. *Lilium brownii* also continues to be grown as a crop over a wide area. Both these two lilies were introduced into Britain from cultivation in China. They arrived at Kew together in 1804, having been obtained from a nursery in Guangzhou by William Kerr, and sent to Britain on the East Indiaman *Henry Addington*. *Lilium longiflorum* may have come by a similar route a decade or so later, though there is no reliable record of its means of introduction. *Lilium concolor* almost certainly also arrived on a British ship from Guangzhou, for it reached the garden in Paddington of the Hon. Charles Greville, one of the founders of the Horticultural Society, before 1806. When Salisbury described it in that year, its country of origin was not known, but soon afterwards it was ascertained to be China. Both the forms with plain and with spotted flowers were grown by 1809, and both were again introduced from Shanghai by Robert Fortune in the 1840s. He obtained a large proportion of the plants he introduced from Chinese nursery gardens. Chinese gardeners thus made a direct contribution to the enrichment of western gardens.

Subsequent exploration by western plant-hunters of the wilder parts of China, particularly in the south-west, resulted in the discovery and introduction (to both Europe and America) of most of the other species of Chinese lilies. Though some have since been lost to cultivation, only a few have never been grown at all. Despite difficulties with their cultivation and fluctuations in their popularity, Chinese lilies have made a considerable contribution to the beauty of western gardens, whether directly or through their hybrid offspring, just as their ancestors enhanced the gardens of China. The long history of their appreciation and cultivation continues to grow ever longer and richer.

CHAPTER FOUR

The taxonomy of Chinese lilies

Chinese botanists face a major difficulty in trying to assess the status of many of their native plant species. Because most of the early scientific collecting of plant specimens in China was undertaken by westerners, the type specimens of a very high proportion of Chinese species are located in western herbaria. This is particularly true of plants of ornamental value (including, of course, lilies), which were especially sought after for foreign gardens. It is only necessary to think of the collections of Robert Fortune, Père Delavay, Augustine Henry, E. H. Wilson, George Forrest, Frank Kingdon Ward and Dr Joseph Rock (and there were many others), to realize just how much material was brought out of China to be deposited in foreign herbaria. It is now not easy for Chinese botanists to consult this important material. Conversely, westerners have a great mass of old pressed specimens to work on, but very few more recent collections, and find it very difficult indeed to undertake any further studies in the field. Permission for such work is rarely granted to foreigners by the Chinese government, and access to most of China is still restricted, even after recent relaxations. Under these conditions, it is hardly surprising that the taxonomic problems of many groups of Chinese plants are still not very satisfactorily resolved.

The treatment which the *Flora Reipublicae Popularis Sinicae* accords to the four genera considered in this book raises a number of such problems, and disagrees in several respects with that commonly accepted by western botanists. *Lilium apertum* Franch., for example, was long ago transferred to the genus *Nomocharis* (by E. H. Wilson in 1925), and no westerner has wished to alter this since. It was therefore rather surprising to find both this species and *Nomocharis soluenensis* Balf. f. replaced in the genus *Lilium* by Liang Sung-yun. There is also disagreement regarding the concepts of several species and varieties, some being merged together and others maintained as distinct where westerners have usually taken the opposite view. Finally, the sectional divisions of the genus *Lilium* agree rather badly with the most recent western opinions on the relationships between species within the genus. Only four sections are maintained, and two of these are made to contain a somewhat disparate assortment of species. These problems require very careful consideration. In order to resolve them, it is necessary to arrive at a reasonably clear view of the distinctions between the Chinese lily species and their relationships to each other.

The four genera are certainly very closely related, so closely in fact that almost all of their species have at one time or another been included within the single genus *Lilium*. *Cardiocrinum* is now ranked as a distinct genus on the basis of its long-stalked, heart-shaped or ovate leaves and monocarpic bulb formed of the swollen bases of the stalks of the basal leaves. It must have diverged from the genus *Lilium* at quite an early evolutionary stage. The genus *Notholirion* exhibits even more primitive characteristics, its monocarpic bulb being formed of the bases of its strap-shaped leaves, and its flowers being rather small and openly campanulate. The species of this genus are probably closest in appearance to the ancestral species of all the lilies. In the genus *Lilium* basal leaves have virtually disappeared, and the bulbs are perennial (even if sometimes short-lived). The flowers have developed in several different ways, most extremely into the strongly revolute 'martagon' type. Even after the exclusion from the genus of the species of the other three genera, *Lilium* remains very heterogeneous, with wide variation among its component species. The genus *Nomocharis* is clearly an extreme development of *Lilium*, and in fact the dividing line between these two genera is indistinct and can only be rather arbitrarily drawn. The only really distinctive feature of *Nomocharis* as it is now generally understood lies in the form of the nectaries. In *Nomocharis*, those of the inner perianth segments are highly developed, while on the outer segments they are lacking or rudimentary: in *Lilium* they are developed at least to some extent on both. The basal protuberances on each side of the nectarial furrow are also diagnostic, for *Nomocharis* has strongly developed swellings that are often ridged or crested, and are always deeply coloured. Liang Sung-yun gives the following criterion for distinguishing species of the genus *Nomocharis*:

> We have included those species with fleshy cushion-shaped swellings at the base of the inner perianth segments in this genus, all those without such swellings have been transferred to the genus *Lilium*.

This would seem to be in agreement with the opinion of J. R. Sealy, whose treatments of the genus *Nomocharis* have been generally accepted in the west.

As this is so, it seems strange that Liang Sung-yun should have thought it necessary to transfer *NN. aperta* and *saluenensis* to the genus *Lilium*. It is true that the form of the nectarial processes in these two species is different from that found in the other species of *Nomocharis*, being more or less smooth swellings which are not ridged or crested. Nevertheless, the swellings are definitely present in normally developed plants. The reason for this treatment would seem to be that Miss Liang has separated specimens of *N. aperta* which have weakly-developed nectaries from those in which they are strongly developed, applying the name *Nomocharis forrestii* Balf. f. to the latter. It is, however, quite common in *N. aperta* (sensu lato) for the gynoecium to be reduced or absent, and the nectaries are often correspondingly variable. But this is merely a question of the robustness of the specimen, and continuous variation occurs. I therefore do not think it is justifiable to separate *Nomocharis forrestii* from *Nomocharis aperta*. This being so, there can be no doubt that the species belongs with the genus *Nomocharis*. *N. saluenensis* is so close to *N. aperta* (it was originally described by Franchet as var. *thibeticum* of his *Lilium apertum*) that it must certainly be placed in the same genus. The species descriptions in Chapter 5 have been rearranged accordingly.

The description and line-drawing of *Nomocharis forrestii* in the *Flora RPS* both indicate that the petals are more heavily spotted and blotched than has been thought usual in *Nomocharis aperta*. The amount of maculation in species of this genus is, however, commonly very variable, and is not necessarily taxonomically significant. But it is interesting that one of the characteristics of *Nomocharis synaptica* Sealy is that its flowers are heavily blotched. It is also curious that this latter species, collected by Kingdon Ward near the Assam-Tibet border on just two occasions (in 1928 and 1950), appeared in cultivation in the garden of Major Knox Finlay at Keillour in Scotland (in 1964) without any connection with its discoverer being traceable. It has been speculated that plants may have been grown at the Royal Botanic Garden, Edinburgh, from the very small amount of seed (two capsules) which Kingdon Ward was able to collect in 1928, and that from there they may have passed to Keillour. Although this is certainly possible, it has never been confirmed. I am inclined to suspect that the Keillour plants were in reality derived from collections made in China (or perhaps Burma), which had at first been considered to be typical *Nomocharis aperta*. Indeed, Kingdon Ward himself originally identified his Assam collections (nos. 8399 and 19601) as that species. *N. synaptica* was not described until 1950. Since it can only be separated from *N. aperta* by reason of its ridged rather than smooth nectarial swellings, and since there is in any case considerable variation in the development of the nectarial processes within *N. aperta*, I doubt whether these two species should be maintained as distinct. Probably *N. synaptica* is entitled to no more than subspecific rank, on the basis of its slight morphological differences and geographic isolation. But if it should indeed prove that plants agreeing with *N. synaptica* occur within the range of *N. aperta* further to the east, then no more than varietal status would be appropriate. Further studies of wild populations are needed to resolve this question.

In other respects the treatment of the species of *Nomocharis* in the *Flora RPS* agrees with western opinion of the time, except that *Nomocharis farreri* (W. E. Evans) Harrow was reassigned its original status as a variety of *N. pardanthina* Franchet. Since its publication, however, a new revision of the genus by J. R. Sealy has appeared, which not only maintains the specific status of *N. farreri*, but also merges *N. mairei* Lévl. with *N. pardanthina* Franch. In addition, Sealy expresses the opinion that:

> *N. farreri* could well be a . . . variant of *N. meleagrina*. And *N. basilissa*, on account of its general resemblance to *N. farreri*, would likewise have to be associated with *N. meleagrina*.

It is clear that much more study of the status of all the *Nomocharis* species is needed. Possibly the total number of species in the genus, now considered by Sealy to be seven, should eventually be reduced to no more than four. These would be *N. saluenensis*, as understood by Sealy; *N. aperta*, with which *N. synaptica* would be merged; *N. pardanthina* of Sealy (including *N. mairei*); and *N. meleagrina*, within which *NN. farreri* and *basilissa* would probably retain subspecific or varietal status. It is likely that *Nomocharis* evolved from *Lilium* rather recently, and that it is still in an active evolutionary state. It is for this reason that a wide range of variation occurs, which may well be found to be so continuous as to link many of the presently recognized species. It will not be possible to arrive at a completely satisfactory arrangement for the genus until the variation within wild populations has been studied much more extensively than at present.

Problems concerning the species of the other three genera are happily not very numerous. There are, in fact, none at all in the genera *Notholirion* and *Cardiocrinum* (the only difference from previous treatment of these two genera is that the two varieties of *Cardiocrinum giganteum* are merged). But in the genus *Lilium* a small number of species commonly considered distinct have been reduced to synonymy or varietal rank.

What is probably the most startling of these changes involves the lilies that have been known as *Lilium nepalense* D. Don and, since its nomenclature was revised by Woodcock and Stearn in 1950, *Lilium primulinum* Baker. This includes the lilies to which for a long time the name *Lilium ochraceum* Franchet was applied. Several other synonyms are involved, but these have not been commonly used. In fact, a total of about a dozen specific epithets has been given to lilies of this group (many of them by Léveillé), which is some indication of the taxonomic difficulties which they present.

Lilium nepalense D. Don was first described in 1821 from specimens collected north of Kathmandu in Nepal. Very similar lilies have since been found in many areas of the Himalaya from Kumaon in Uttar Pradesh to just beyond the eastern border of Bhutan. There has never been any major problem with the nomenclature of the plants from this area.

In China and Burma in the late nineteenth century a number of collections were made of lilies which in various respects resembled typical *L. nepalense*. Those from north-west Yunnan and adjacent parts of Sichuan were named *Lilium ochraceum* by Franchet in 1892. Earlier in the same year, the name of *Lilium primulinum* was given by Baker to a yellow-flowered lily from northern Burma. But in 1888 a plant similar to this one, though with considerable purplish coloration on the flower, had been exhibited in

London by the nurserymen Hugh Low and Company, and had been identified by Baker as *L. nepalense*. It originated from Upper Burma.

There was subsequently something of a nomenclatural tangle. The name *L. nepalense* came to be used for most of the lilies of this group whether from Burma or China. Typical plants from the central Himalaya were very rarely collected and therefore little known, for Nepal and Bhutan were at the time virtually inaccessible to foreigners. This made it very difficult for valid comparisons to be made between the Chinese, Burmese and Nepalese plants.

In 1922 Professor W. Wright Smith produced a revision of the nomenclature applied to these lilies. He joined *L. primulinum* Baker to *L. nepalense* D. Don, as var. *primulinum*. To the similar lily from Upper Burma of which a plant had been shown in 1888, which differed in the heavy brownish-purple coloration in the throat of its flowers, he gave the new name *L. nepalense* var. *burmanicum*. Franchet's *Lilium ochraceum* was maintained as a separate species. He thus recognised two distinct species, one of which had three varieties.

This treatment did not go unchallenged for long. Writing in 1925, E. H. Wilson stated his opinions as follows:

> I am in complete agreement with what he [W. W. Smith] says of this Lily's marked variation in response to different climatic conditions but I can find no characters on which to separate the Yunnan and Burmah forms into different varieties, much less species. At one time I accepted Baker's identification of the Burmah and Nepal Lilies as belonging to one species with which I thought Franchet's species should be united. The advent of good material of *L. nepalense* D. Don from the neighbourhood of Almorah has convinced me that this species is distinct from that of Burmah and western China which Franchet named *L. ochraceum*. Finally, my investigations have led me to conclude that the Himalayan Lily is a true Leucolirion, and that of Burmah and Yunnan a true Martagon.

He therefore separated the two varieties *primulinum* and *burmanicum* from *L. nepalense*, and joined them to *L. ochraceum*, maintaining only var. *primulinum* as a distinct variety. But it is interesting that he was at first inclined to merge all three species, yet subsequently placed the two he finally recognized in separate sections of the genus. His section Leucolirion corresponds largely with Baranova's section Regalia, containing all the Chinese and Japanese trumpet-lilies. This suggests that his knowledge of *L. nepalense* was far from adequate. Indeed, he himself admitted as much. *L. nepalense*, he wrote:

> is to-day one of the least known of all Lilies. . . . The plant is unknown to me in the living state and I have not been able to find any description of the bulb or fruit. . . . So far as I can discover there is no evidence to prove that Don's species exhibits any variation from the typical funnel-form flower characteristic of the section Leucolirion.

There is no question but that he was entirely wrong to consider that *L. nepalense* had the 'typical' flower-form of the trumpet-lilies. It is of course notoriously difficult to be sure of such a characteristic from pressed specimens, in which the shape and disposition of the perianth is commonly greatly distorted. In fact, *L. nepalense* has flowers which are closest in form to those of *Lilium monadelphum* and its near relatives, being pendulous or nodding with a moderately long tube (but by no means as long as in the lilies of section Regalia), and perianth segments strongly recurved or revolute in their upper half. Comber placed *L. nepalense* close to *L. ochraceum* in his section Sinomartagon.

Subsequently, Woodcock and Stearn pointed out that the name *Lilium primulinum* Baker had several months' priority over *L. ochraceum* Franchet, so that the latter had to be regarded as a synonym of the former, rather than vice versa. Otherwise they agreed with E. H. Wilson's treatment, except that they revived the variety *burmanicum*. Their opinion has been commonly accepted in the west.

The reasons which they gave for continuing to regard *L. nepalense* as distinct from *L. primulinum*, though perhaps sounder than E. H. Wilson's, are nevertheless open to doubt. I have examined a large number of herbarium specimens of these lilies, and have found *L. nepalense* from the central Himalaya to be much more variable than they seem to have believed. Certain forms from Nepal with much smaller flowers than is common in plants from the area appear very close to Franchet's *L. ochraceum* from north-west Yunnan. There are also some specimens from southern Yunnan, especially from near Mengzi (Mengtze), which are very similar to the common large-flowered forms from the central Himalaya. The only really constant differences lie in the heights of the stems and the width and veining of the leaves, though even among specimens from Nepal only there is considerable variation in these characters. For these reasons, I think it probable that the treatment of these lilies in the *Flora RPS* is correct, though possibly the varieties there recognized may merit subspecific rank if their differences in morphology can be shown to correlate with geographic distribution.

The other changes in specific status made in the Chinese *Flora* are much more straightforward. Firstly, *Lilium euxanthum* (W. Smith and W. E. Evans) Sealy has been merged with *Lilium nanum* Klotzsch var. *flavidum* (Rendle) Sealy. At the time when *L. euxanthum* was first described (as *Nomocharis euxantha*) in 1925, *Fritillaria flavida* Rendle had not yet been made a variety of *L. nanum*. The original specific differentiation of the two lilies was therefore based on comparison of the yellow-flowered plants assigned to *L. euxanthum* with the purplish-flowered type variety of *L. nanum*. Recently it has been suggested by some authorities that the two varieties of *L. nanum* should be considered distinct species. Whether or not this is justifiable, it is clear that *L. euxanthum* is closer to *L. nanum* var. *flavidum* than the latter is to *L. nanum* var. *nanum*. The union of the two yellow- to white-flowered lilies must therefore, I think, be accepted.

The last merger of species is that of *Lilium lankongense* Franchet with *Lilium duchartrei* Franchet. It has always been recognized that these two lilies were very closely related, and doubts have often been expressed as to whether they could really be considered separate species. Wilson stated that:

> The position of *L. lankongense* as a species distinct from Duchartre's Lily depends on the constancy of its truly racemose as opposed to the umbellate or subumbellate inflorescence of the latter.

The forms of these lilies grown in western gardens are undoubtedly constant in the characters of their inflorescences, and are readily distinguished even by untrained eyes. It is, however, highly likely that wild populations show much more variation. Liang Sung-yun justifies her merger of the species by saying:

> Examination of the specimens kept at the Botanical Institute of Academia Sinica shows that the inflorescence varies greatly, from a solitary flower to a raceme or more or less an umbel, sometimes apparently being umbelliform, but with pedicels of unequal lengths, sometimes with a leaf-like bract on the middle of the pedicel, while the inflorescence is really still a raceme; as for the other features, such as whether the leaves are crowded or not, the prominence of the nerves on the undersides of the leaves and the colour of the flowers, these are all variable, with intermediate and overlapping forms.

My own examination of herbarium specimens has convinced me that this opinion should be accepted. It must be remembered that the plants now in cultivation in our gardens as *L. duchartrei* derive more or less exclusively from collections made by Farrer and Purdom in southern Gansu, at the northern limit of the lily's distribution, while conversely our stocks of *L. lankongense* originally came from north-west Yunnan, hundreds of miles to the south. They should therefore be expected to represent the most extremely differentiated forms of this lily, and must not be assumed to exemplify the range of variation found in nature.

One new *Lilium* species is published in the account of the genus in the *Flora RPS*. This is *Lilium xanthellum* Wang et Tang, described as very similar to *Lilium fargesii* Franchet, but with a very much larger bulb and yellow flowers. The difference in flower coloration would hardly be sufficient on its own to justify the creation of a new species, but if the very large bulb is a constant attribute then this lily is clearly distinct.

There are a couple of lilies recorded from China which are not mentioned in the *Flora RPS*. One is *Lilium medeoloides* A. Gray, which has frequently been said to occur in Zhejiang province. This seems to be based on only one collection, however, and it must be strongly doubted whether this was correctly identified. If it was, then it may well be that it represents a cultivated plant not native to China. This species has been grown in Japanese gardens for several centuries, and could easily have reached Chinese gardens from Japan. Several other lilies are known to have done so.

Lilium pyi Lévl. is also omitted. It is very probable, however, that this name should be referred to some other known Chinese species. As no extant specimen of the lily is known, its correct status cannot now be determined.

It is very noticeable that quite a number of varieties do not appear in the Chinese *Flora*. These include such well-known lilies as *Lilium davidii* Duchartre var. *willmottiae* (Wilson) Raffill, and the several varieties of the Tiger Lily. Perhaps their omission indicates that they are not considered distinct enough to be maintained, but if so it is strange that they are not listed as synonyms.

Apart from the differences discussed above, the concepts of the Chinese species of *Lilium* are generally familiar. Miss Liang's overall treatment of the genus is rather conservative, with little in the way of major revisions. Her division of the genus into sections, however, does not seem very satisfactory. *Lilium concolor* Salisb. and *Lilium dauricum* Ker-Gawl. do not associate well with *L. nanum* Klotzsch and the other species of section Lophophorum (Bur. et Franch.) Wang et Tang, for example. The basic problem is that the genus (as represented in China) is divided into only four sections, which is far too few, thus necessitating the inclusion in at least some of the sections of anomalous species.

In fact a completely satisfactory division of the genus *Lilium* into sections has never been proposed. The best arrangement yet devised is that of Comber, published in the *RHS Lily Year Book* of 1949. He recognized that the form of the flower had been relied on far too heavily in the past as an indicator of relationships within the genus, and that many other factors had to be taken into account. His division into seven sections nevertheless seems to have been far too much influenced by geographical factors, for all the European species (except *Lilium martagon*) are placed together in one section, while all the North American species comprise another. This seems to me to ignore completely the very great morphological differences between, for example, *Lilium candidum* and *Lilium bulbiferum*, or *Lilium catesbaei* and *Lilium pardalinum*. His treatment of the Asian species is nevertheless generally satisfactory.

Although Woodcock and Stearn in 1950 did not attempt a full revision of the sections of *Lilium*, they did present a 'provisional tabulation', which classed the species into fifteen groups, a few of which were further divided into subgroups. Their arrangement has much to be said for it, but in several instances seems to introduce unnecessary divisions, as for example between *Lilium bulbiferum* and *Lilium dauricum*.

There seems to have been no serious attempt to make a full revision of the sectional divisions since the publication of Comber's classification. P. M. Synge, in his major work *Lilies* published by Batsford in 1980, makes no mention at all of subgeneric divisions of *Lilium*, and prefers a geographic and alphabetic arrangement of species. There have, however, been some contributions towards a more natural sectional classification, notably by the Russian botanist M. Baranova, who in 1970 published a new section Regalia for the trumpet-lilies. The horticultural classification drawn up in 1963 by the Lily Committee of the Royal Horticultural Society has no taxonomic status but is at least partly based on natural relationships.

In many genera recent developments in the study of chromosomes has assisted considerably in arriving at a satisfactory natural classification. Unfortunately this has not been the case with *Lilium*. The chromosomes of the various species of the genus so far studied (and including those of *Cardiocrinum giganteum*) have proved to be so similar as to be of no use in deciding relationships between the species. As far as is known, all *Lilium* species possess a regular diploid chromosome number of 2n = 24, except for some forms of *LL. lancifolium* and *japonicum* which are triploids with 2n = 36. There is not uncommonly intraspecific variation in karyotypes, and it is impossible to discern any clear specific groupings in chromosome morphology. Cytological studies have therefore been of no help so far in distinguishing related groups of *Lilium* species.

After considering available evidence and the opinions of previous authors, I have drawn up my own classification of the genus. This proposes a total of 13 sections, eight of which are composed of or include species which occur in China. Two of these Chinese sections are new. For the purposes of this book, I shall here propose formal taxonomic status only for the sections involving Chinese species, making informal suggestions as to the remainder.

Infrageneric classification of the genus *Lilium*

(1) Section **Lilium**

Type and only species: *L. candidum* L.

A very distinct section, with basal leaves and openly funnel- or bowl-shaped flowers arranged in a raceme.

(2) Section **Asteridium** S. G. Haw, sect. nov. in Appendix 1.

Type and only species: *L. concolor* Salisb.

This species is so difficult to associate satisfactorily with any of the other species of the genus that it seems best to give it its own section. It has in the past been placed in several very different sections, including Pseudolirium by E. H. Wilson, Sinomartagon by Comber, and Lophophorum by Wang and Tang.

(3) Section **Pseudolirium** Endlicher, Gen. Pl., 141. 1837, pro subgen.; Wilson, *Lil. E. Asia*, 50. 1925; emend.

Type species: *L. catesbaei* Walter.

Other species: *L. philadelphicum* L., *L. dauricum* Ker-Gawl., *L. maculatum* Thunb., *L. bulbiferum* L.

This seems to me to be a quite natural grouping, all the species having erect flowers with perianth segments distinctly narrowed into claws at the base. One species has whorled foliage, while in the others the leaves are scattered; but there is a tendency in all the species for the leaves immediately below the umbellate inflorescence to form a whorl. Germination in all these lilies is hypogeal, so far as I am able to ascertain, though there are conflicting statements about this in the literature.

(4) Section **Regalia** Baranova in *Nov. Sist. Vyssh. Rast.* 8: 91–94. 1971.
Type species: *L. regale* Wilson.
Other species: *L. sargentiae* Wils., *L. leucanthum* Baker, *L. sulphureum* Baker, *L. formosanum* Wallace, *L. longiflorum* Thunb., *L. wallichianum* Schult. f., *L. philippinense* Baker, *L. brownii* Brown ex Miellez, *L. japonicum* Thunb., *L. rubellum* Baker, *L. nobilissimum* Makino, *L. alexandrae* (Wallace) Coutts.

This group agrees fairly closely with Wilson's section Leucolirion, but as he included *L. candidum* in the section, his name cannot be applied to this section once *L. candidum* has been excluded. I have therefore followed M. V. Baranova and applied the name Regalia to the section. It seems to me to be reasonably homogeneous, but the species from Japan (with the exception of *L. longiflorum*) approach the lilies of section Archelirion in certain respects, particularly in having petiolate leaves and hypogeal germination.

(5) Section **Lophophorum** (Bur. et Franch.) Wang et Tang; emend. – *Fritillaria* section **Lophophora** Bur. et Franch. in *Journ. de Bot.* 5: 154. 1891
Type species: *L. lophophorum* (Bur. et Franch.) Franch.
Other species: *L. nanum* Klotzsch, *L. oxypetalum* (Royle) Baker, *L. paradoxum* Stearn, *L. amoenum* Wilson ex Sealy, *L. sempervivoideum* Lévl., *L. sherriffiae* Stearn, *L. georgei* (W. E. Evans) Sealy, *L. souliei* (Franch.) Sealy, *L. bakerianum* Coll. et Hemsl., *L. henrici* Franch., *L. mackliniae* Sealy.

This is a somewhat heterogeneous section, but all the species it includes are clearly closely related to the lilies of genus *Nomocharis*, and many have in the past been included within that genus. The flowers of all the species are more or less nodding, and campanulate to openly campanulate, with strongly developed nectaries especially on the inner perianth segments. I can see no convincing way to further divide this group.

(6) Section **Archelirion** Baker in *Gard. Chron.*, 1871; 104. 1871; emend.
Type species: *L. auratum* Lindley.
Other species: *L. speciosum* Thunb.

Although these two species appear superficially rather different, it is really only the form of the perianth which gives this impression. In all other respects they are closely similar, and seem to me to form a natural group. They are clearly closely related to the lilies of section Regalia, and probably also to those of section Dimorphophyllum. These relationships would seem to be confirmed by the hybrids which have been produced between species of the three sections.

(7) Section **Dimorphophyllum** S. G. Haw, sect nov. in Appendix 1.
Type species: *L. henryi* Baker.
Other species: *L. rosthornii* Diels.

These species have in the past usually been associated with those of Comber's section Sinomartagon. This is largely because of the form of their flowers, however, and in most other characteristics they are distinct. In particular, their dimorphous foliation is a unique feature, which clearly sets them apart. The fact that *L. henryi* has hybridized with *L. speciosum* and *L. sargentiae* may indicate a closer relationship with the sections Archelirion and Regalia than with section Sinomartagon.

(8) Section **Sinomartagon** Comber in *Lily Year Book* 13: 101. 1949; emend.
Type species: *L. davidii* Duchartre.
Other species: *L. lancifolium* Thunb., *L. leichtlinii* Hook. f., *L. papilliferum* Franch., *L. duchartrei* Franch., *L. nepalense* D. Don, *L. poilanei* Gagnepain, *L. arboricola* Stearn, *L. taliense* Franch., *L. stewartianum* Balf. f. et W. W. Sm., *L. wardii* Stapf ex Stearn, *L. fargesii* Franch., *L. xanthellum* Wang et Tang, *L. callosum* Sieb. et Zucc., *L. cernuum* Komar., *L. pumilum* Delile, *L. amabile* Palibin, *L. pyrenaicum* Gouan, *L. ciliatum* P. H. Davis, *L. chalcedonicum* L., *L. pomponium* L.

This is a large and rather heterogeneous section, which nevertheless is difficult to divide further. I have reduced the number of Asian species which Comber originally included in this section, but have then added to it all the European lilies with 'martagon'-form flowers (with the exception of *L. martagon* itself). I cannot see any major point on which these European species with scattered leaves and strongly revolute perianth segments differ from the Asian species with similar morphological features. Comber was quite wrong to state that the lilies of his section Liriotypus, of which these European 'martagons' formed the majority, were not or rarely stem-rooting.

(9) Section **Martagon** Duby, in *Bot. Gall.* 1: 462. 1828; emend. Comber in *Lily Year Book* 13: 98–100. 1949.
Type species: *L. martagon* L.
Other species: *L. hansonii* Leichtlin, *L. distichum* Nakai, *L. medeoloides* A. Gray, *L. tsingtauense* Gilg.

Despite the great variation in the form and disposition of the flower in this section, there can be no doubt that the species within it are very closely related, all having whorled leaves and hypogeal germination. They clearly form a link between the European and Asian species of *Lilium* and those from North America (apart from *LL. catesbaei* and *philadelphicum*).

Other species

The above nine sections include the type and all the Chinese species of the genus, so that those remaining do not fall within the scope of this book. I have necessarily had to consider the genus as a whole when devising this classification, however, and would suggest that the rest of the species may be grouped as follows:

(a) *LL. monadelphum* M. Bieberstein, *kesselringianum* Mischenko, *ledebourii* (Baker) Boissier, *rhodopaeum* Delipavlov, *polyphyllum* D. Don.

These lilies are obviously close to those of section Sinomartagon, but have hypogeal germination and differ from all but the large-flowered varieties of *L. nepalense* in the form of their flowers.

(b) *LL. columbianum* Hanson, *washingtonianum* Kellog, *humboldtii* Roezl et Leichtlin, *rubescens* S. Watson, *bolanderi* S. Watson, *kelloggii* Purdy.

The North American lilies of this group are characterized mainly by the form of the bulb, which grows in one direction only in a manner which has been described as sub-rhizomatous. It therefore has an uneven appearance, with the stem not shooting from the centre, as is normal in the European and Asian species.

(c) *LL. pardalinum* Kellogg, *occidentale* Purdy, *kelleyanum* Lemmon, *nevadense* Eastwood, *pitkinense* Beane et Vollmer, *vollmeri* Eastwood, *wigginsii* Beane et Vollmer, *parryi* S. Watson, *maritimum* Kellogg, *parvum* Kellogg.

These are the American Pacific Coast lilies, with more or less rhizomatous bulbs, and generally enjoying plenty of moisture.

(d) *LL. canadense* L., *grayi* S. Watson, *michauxii* Poiret, *superbum* L., *michiganense* Farwell, *iridollae* M. G. Henry.

In these lilies the bulb is annual, with a new bulb forming each year at the end of a short stolon (which is not a stoloniform stem as in some Asian species).

My sectional divisions of *Lilium* compare with those of the *Flora RPS* as follows:

Section Lilium of Liang is composed of the same Chinese species as are included in section Regalia, but if *L. candidum* is excluded then Baranova's name must be accepted.

Section Lophophorum of Liang is similar to my section Lophophorum, but Liang includes *LL. concolor* and *dauricum*. I have placed the former in its own new section Asteridium, and the latter in section Pseudólirium.

Section Sinomartagon of Liang includes all the Chinese species which I consider should belong here, but contains several more which I would place elsewhere. These are *LL. apertum* and *saluenense*, which I have replaced in the genus *Nomocharis*, following Sealy; *L. speciosum*, which I have transferred to section Archelirion; and *LL. henryi* and *rosthornii*, for which I have created the new section Dimorphophyllum.

Section Martagon contains the same Chinese species in both treatments.

CHAPTER FIVE

Descriptions of the Chinese lily species

The following descriptions and keys are translated from the text by Miss Liang Sung-yun of the Botanical Institute of Academia Sinica, Beijing, published in the *Flora Reipublicae Popularis Sinicae*, Vol. 14, pp. 116–66. Certain amendments have been made. In particular, the species of the genus *Lilium* have been rearranged into the sections which I have proposed in the previous chapter. The numbering of their order in the original text is given in brackets at the head of each species description, after the new sequence number. The genus *Nomocharis* has also been rearranged in the light of the recent revision of the genus by J. R. Sealy. Two species placed in *Lilium* by Miss Liang have been transferred back to *Nomocharis*, and some species concepts have been altered. The Keys to these two genera have needed a great deal of revision as a result of these changes. Following each translated description, I have added various comments of my own, marked at their beginning with a † sign. Any substantial alterations to the original description are noted and explained in these comments.

Lilium L.

L., *Sp. Pl. ed.* 1, 302. 1753; et *Gen. Pl. ed.* 5, 143. 1754.

Bulb commonly ovoid or subglobose; scales numerous, fleshy, ovate or lanceolate, with or without nodes, usually white or yellow. Stem terete, papillose or not, sometimes streaked with purple. Leaves usually scattered, less often whorled, lanceolate, oblong-lanceolate, oblong-oblanceolate, elliptic or linear, sessile or with short petioles, entire or with papillose margins. Flowers solitary or arranged in a raceme, less often more or less umbellate or corymbose; bracts similar to the leaves but smaller; perianth often brightly-coloured, sometimes fragrant; perianth segments six, in two whorls, separate, frequently more or less connivent and forming a trumpet- or bell-shape, or strongly revolute, usually lanceolate or spathulate, with nectaries at the base, the nectaries with or without papillae on each side, and sometimes also with cristate or fimbriate projections; stamens six, filaments awl-shaped, hairy or smooth, anthers ellipsoid, dorsally attached, versatile; ovary cylindric; style generally relatively slender, stigma swollen, trilobed. Capsule oblong, dehiscing loculicidally. Seeds numerous, flattened, winged all round the circumference.

About 90 species, distributed throughout the north temperate zone. China has 37 species, the type varieties of three of which do not occur in China, distributed evenly across the north and the south, but particularly numerous in south-west and central China. The bulb contains starch and can be eaten, some species are used medicinally, and the flowers of some species contain essential oil, which can be extracted for use in perfume.

Key to the species of the genus *Lilium*

1. Leaves scattered.
   2. Flowers erect.
      3. Bases of the leaves not woolly; style shorter than the ovary; perianth segments 2.2–3.5cm long; papillae on each side of the nectaries not deep purple (Jilin, Hebei, Shandong, Henan, Shanxi, Shaanxi, Yunnan) .... 1. *L. concolor* Salisb.
      3. Bases of the leaves with a cluster of woolly hairs; style at least twice as long as the ovary; perianth segments 7–9cm long; papillae near the nectaries dark purple-red (Heilongjiang, Liaoning, Jilin, Inner Mongolia, Hebei)....................... 2. *L. dauricum* Ker-Gawl.
   2. Flowers more or less horizontal or nodding.
      4. Flowers more or less horizontal; perianth trumpet-shaped, white or yellowish inside, unspotted, segments recurved at the tips; stamens curved upwards near their tips and more or less exserted.
         5. Nectaries papillose; leaf-axils not bulbiliferous; leaves oblanceolate or obovate to lanceolate, occasionally linear, (0.6–) 1–2cm

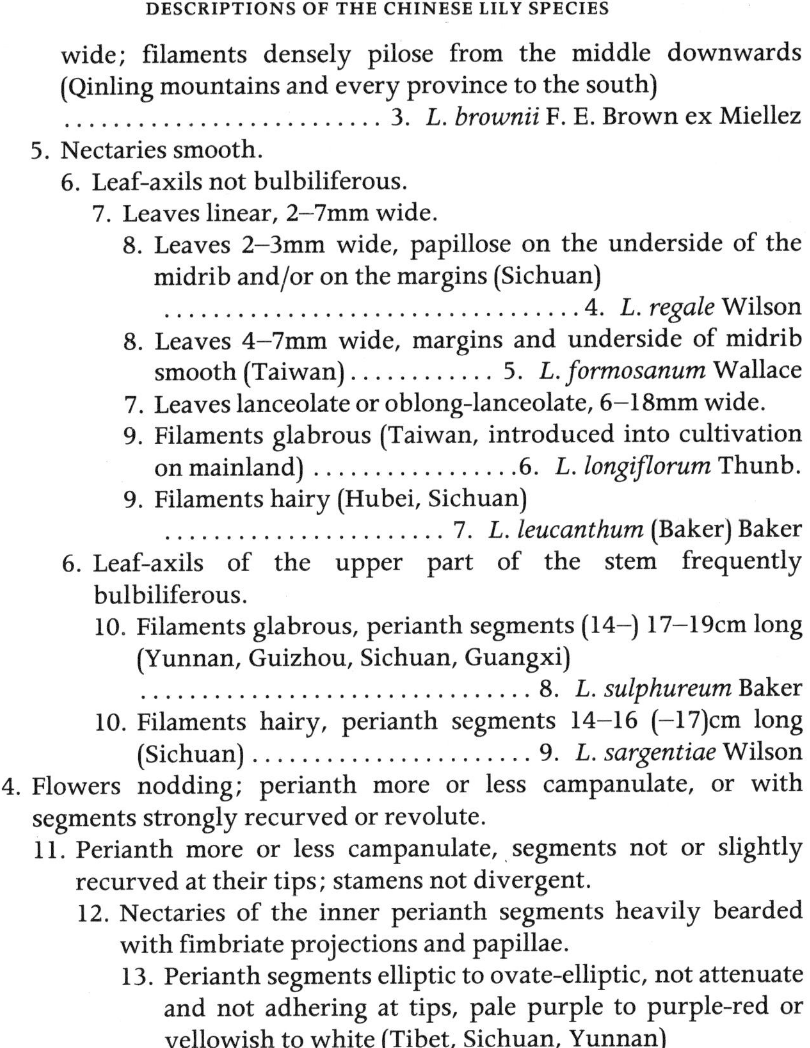

wide; filaments densely pilose from the middle downwards (Qinling mountains and every province to the south)
.... 3. *L. brownii* F. E. Brown ex Miellez

5. Nectaries smooth.
  6. Leaf-axils not bulbiliferous.
    7. Leaves linear, 2–7mm wide.
      8. Leaves 2–3mm wide, papillose on the underside of the midrib and/or on the margins (Sichuan)
      .... 4. *L. regale* Wilson
      8. Leaves 4–7mm wide, margins and underside of midrib smooth (Taiwan) .... 5. *L. formosanum* Wallace
    7. Leaves lanceolate or oblong-lanceolate, 6–18mm wide.
      9. Filaments glabrous (Taiwan, introduced into cultivation on mainland) .... 6. *L. longiflorum* Thunb.
      9. Filaments hairy (Hubei, Sichuan)
      .... 7. *L. leucanthum* (Baker) Baker
  6. Leaf-axils of the upper part of the stem frequently bulbiliferous.
    10. Filaments glabrous, perianth segments (14–) 17–19cm long (Yunnan, Guizhou, Sichuan, Guangxi)
    .... 8. *L. sulphureum* Baker
    10. Filaments hairy, perianth segments 14–16 (–17)cm long (Sichuan) .... 9. *L. sargentiae* Wilson

4. Flowers nodding; perianth more or less campanulate, or with segments strongly recurved or revolute.
  11. Perianth more or less campanulate, segments not or slightly recurved at their tips; stamens not divergent.
    12. Nectaries of the inner perianth segments heavily bearded with fimbriate projections and papillae.
      13. Perianth segments elliptic to ovate-elliptic, not attenuate and not adhering at tips, pale purple to purple-red or yellowish to white (Tibet, Sichuan, Yunnan)
      .... 11. *L. nanum* Klotzsch
      13. Perianth segments lanceolate to ovate-lanceolate, 4.5–5.7cm long, 0.9–1.6cm wide, attenuate, often adhering at the tips, yellowish (Sichuan, Yunnan, Tibet)
      .... 12. *L. lophophorum* (Bur. et Franch.) Franch.
    12. Nectaries of the inner perianth segments not heavily bearded, smooth to lightly papillose.
      14. Leaves 2.5–8cm long; flowers often solitary, or 3–4; perianth segments of a variety of colours, but without any basal blotch.
        15. Stem not papillose; flowers purple-red, unspotted; bulb clearly elongated, rather narrowly ovoid, about twice as tall as broad (Sichuan, Yunnan)

.... 13. *L. souliei* (Franch.) Sealy

15. Stem papillose; flowers of a variety of colours, but usually spotted; bulb sub-ovoid or spherical, roughly as tall as broad.
    16. Margins and undersides of the midribs of the leaves papillose; flowers white, pink or yellow, frequently spotted with purple inside (Yunnan, Sichuan, Guizhou)
    .... 10. *L. bakerianum* Coll. et Hemsley
    16. Leaf-margins not papillose; flowers white or purple-red, finely spotted.
        17. Leaves 16–25, linear; flowers white (Yunnan, Sichuan)......14. *L. sempervivoideum* Lévl.
        17. Leaves 8–12, elliptic or narrowly oblong; flowers purple-red or purple-rose (Yunnan)
        .... 15. *L. amoenum* Wilson ex Sealy

14. Leaves 12–15cm long; flowers often 5–6, occasionally solitary; perianth segments white or suffused with rose-purple, with a deep purple blotch on the inside at the base (Yunnan, Sichuan)..........17. *L. henrici* Franch.

11. Perianth not campanulate, segments strongly recurved or revolute; stamens frequently widely divergent.
    18. Leaves shortly but distinctly petiolate; nectaries strongly papillose and with fimbriate projections on each side.
        19. Leaves all of more or less the same shape; flowers white or tinged with pink; margins of the perianth segments undulate (Anhui, Jiangxi, Zhejiang, Hunan, Guangxi, Taiwan) .18. *L. speciosum* Thunb. var *gloriosoides* Baker
        19. Leaves clearly dimorphous; flowers yellow or orange; perianth segments entire.
            20. Leaves oblong-lanceolate, rounded at the base, 2–2.7cm wide; capsule oblong, 4–4.5cm long, c. 3.5cm broad, brown (Hubei, Jiangxi, Guizhou).
            ....19. *L. henryi* Baker
            20. Leaves linear-lanceolate, 0.8–1cm wide; capsule long-oblong, 5.5–6.5cm long, 1.4–1.8cm broad, olive-green (Sichuan, Hubei, Guizhou)
            .... 20. *L. rosthornii* Diels
    18. Leaves sessile or indistinctly petiolate; nectaries papillose or smooth.
        21. Nectaries smooth.
            22. Flowers yellow or greenish-yellow, throat frequently deep purple, unspotted (Tibet, Yunnan, Sichuan)..................21. *L. nepalense* D. Don

22. Flowers white, pale reddish-purple, pink or greenish-yellow, throat not deep purple, but spotted with purple-red.
    23. Flowers pale reddish-purple or pink; style at least three times as long as the ovary; leaves narrowly lanceolate, with three veins impressed on the upper surface (Tibet) .... 22. *L. wardii* Stapf ex Stearn
    23. Flowers white or greenish-yellow; style equal to or a little longer than the ovary; leaves linear or linear-lanceolate, without three veins impressed on the upper surface.
        24. Flowers solitary, greenish-yellow, with deep red markings (Yunnan)
            23. *L. stewartianum* Balf. f. et W.W. Sm.
        24. Flowers 2–5 (–13) arranged in a raceme, white, with purple markings (Yunnan, Sichuan) . . . 24. *L. taliense* Franch.

21. Nectaries with papillae on each side.
    25. Leaf-axils not bulbiliferous.
        26. Leaves lanceolate to oblong.
            27. Flowers white or pink, with purple spots, perianth segments without fimbriate projections (Sichuan, Yunnan, Tibet, Gansu) ....25. *L. duchartrei* Franch.
            27. Flowers orange-red to red, with purple spots, perianth segments with fimbriate projections (Shaanxi, North China, Manchuria)
                27. *L. leichtlinii* Hook. f. var. *maximowiczii* (Regel) Baker
        26. Leaves linear.
            28. Nectaries papillose on each side, but without cristate projections.
                29. Bracts not thickened at the tip.
                    30. Stem scabrid or papillose; flowers orange to scarlet, spotted with purple (Sichuan, Yunnan, Shaanxi, Gansu, Henan, Shanxi, Hubei) .... 26. *L. davidii* Duchartre
                    30. Stem smooth or puberulous, not scabrid or papillose.
                        31. Flowers bright scarlet, usually unspotted, occasionally lightly spotted (North China, Manch-

uria, North-west China excluding Xinjiang)
.... 29. *L. pumilum* Delile

31. Flowers pale purple-red or rose-pink, lightly to heavily spotted with deep purple (Jilin)
........30. *L. cernuum* Komar.

29. Bracts thickened at the tip; flowers comparatively small; perianth segments red or light red, almost unspotted; style shorter than the ovary (Taiwan, Guangdong, Zhejiang, Jiangsu, Anhui, Henan, Manchuria)
..........31. *L. callosum* Sieb. et Zucc.

28. Nectaries with both papillae and cristate projections on each side.

32. Bulb white, 2–3cm tall, 1.5–2.5cm in diameter.

33. Flowers purple-red (Yunnan, Sichuan, Shanxi)
........32. *L. papilliferum* Franch.

33. Flowers greenish-white, with dense purple-brown spots (Yunnan, Sichuan, Hubei, Shaanxi)
.... 33. *L. fargesii* Franch.

32. Bulb yellow, about 4.5cm tall, 4–5cm in diameter (Sichuan)
......34. *L. xanthellum* Wang et Tang

25. Leaf-axils of the upper part of the stem bulbiliferous; flowers vermilion, with purplish-black spots (East China, Central China, North China, North-west China, Guangxi and Jilin)
....................28. *L. lancifolium* Thunb.

1. Leaves verticillate.

34. Leaves in two or more whorls; flowers campanulate; stamens not divergent (Tibet) ....16. *L. paradoxum* Stearn

34. Leaves usually in one whorl; flowers not campanulate; stamens divergent.

35. Flowers erect, perianth segments spreading, slightly recurved but not revolute, orange, spotted, nectaries not papillose (Shangdong, Anhui)................................ 35. *L. tsingtauense* Gilg

35. Flowers not erect, perianth segments more or less revolute.

36. Bulb scales jointed; leaves obovate-lanceolate to oblong-lanceolate; perianth segments pale vermilion, spotted, nectaries not, or only lightly, papillose (Jilin, Liaoning)

.................................36. *L. distichum* Nakai

36. Bulb scales unjointed; leaves lanceolate to oblong-lanceolate; perianth segments purple-red, spotted, nectaries densely papillose on each side (Xinjiang)
................37. *L. martagon* L. var. *pilosiusculum* Freyn

I. Section Asteridium S. G. Haw, sect. nov, in Appendix 1.

1 (10). **Lilium concolor** Salisb. in *Hook. Parad. Lond.* 1 : t. 47. 1806. – *L. mairei* Lévl. in *Rep. Sp. Nov. Fedde* 11 : 303. 1912.

Bulb ovoid, 2–3.5cm tall, 2–3.5cm in diameter; scales ovate or ovate-lanceolate, 2–2.5 (–3.5)cm long, 1–1.5 (–3)cm wide, white, with roots on the stem above the bulb. Stem 30–50cm tall, occasionally tinged with purple near the base, papillose. Leaves scattered, linear, 3.5–7cm long, 3–6mm wide, 3–7-nerved, margins papillose, glabrous on both surfaces. Flowers 1–5 in sub-umbelliform or racemose arrangement; pedicels 1.2–4.5cm long: perianth erect, stellate, deep red, unspotted, shining; perianth segments oblong-lanceolate, 2.2–4cm long, 4–7cm wide, with papillae on each side of the nectaries; stamens converging towards the centre; filaments 1.8–2cm long, glabrous, anthers long-oblong, *c.* 7mm long; ovary cylindric, 1–1.2cm long, 2.5–3mm broad; style slightly shorter than the ovary, stigma somewhat swollen. Capsule oblong, 3–3.5cm long, *c.* 2.2cm wide.

Flowering period: June–July. Fruiting period: August–September.

Occurs in Henan, Hebei, Shandong, Shanxi, Shaanxi, Jilin, western Hubei and north-east Yunnan (Dongchuan area; it is very rare in Yunnan). It grows in grassy places on mountain slopes, by the sides of roads, and in scrub, at 350–2000 metres above sea level. The bulb contains starch and can be eaten or used to make wine; it may also be used medicinally and is effective as a tonic and strong anti-tussive. The flower contains essential oils, which can be used to make perfume.

var. **pulchellum** (Fisch.) Regel in *Gartenfl.* 25: 354. 1876; Woodc. et Stearn *Lil. World* 201. f. 39. 1950. – *L. pulchellum* Fisch. in Fisch., Mey. et Ave-Lallem. in *Ind. Sem. Hort. Petrop.* 6: 56. 1839.

This variety differs from the type in having spotted perianth segments.

Flowering period: June–July. Fruiting period: August–September.

Occurs in Hebei, Shandong, Shanxi, Inner Mongolia, Liaoning, Heilongjiang and Jilin. It grows in grassy places on sunny slopes, and in moist places in woodland, at 600–2170 metres above sea level. Also distributed in Korea and the USSR.

var. **megalanthum** Wang et Tang in *Flora RPS* 14: 283. 1980.

This variety differs from the type in having wider leaves, 5–10mm wide, and longer perianth segments, 5–5.2cm long, 8–14mm wide, spotted with purple.

Occurs in Jilan (the type specimen was collected near Emu, on a tributary of the Mudan River upstream from Jingpo Lake). It grows in damp meadows at 500 metres above sea level.

† Most previous authors have considered that the form of *Lilium concolor* found in Shandong is distinct from the variety *pulchellum* from Manchuria, Korea and the Amur region. The cited differences between the type variety and var. *pulchellum* are, however, very minor, involving the amount of purple coloration on the stem and the extent to which the leaf-veins are ciliate, characters which in any case are variable. The flora of Shandong has close affinities with that of southern Manchuria and northern Korea, so that it might be expected that the same variety of *Lilium concolor* would occur in both regions. The above treatment, undoubtedly based on examination of far more specimens than have been available to western botanists, therefore seems entirely reasonable.

The relationships of this lily to other *Lilium* species are not at all clear. In the past, it has been placed in various apparently quite distantly related sections of the genus, including both Pseudolirium (by E. H. Wilson) and Sinomartagon (by Comber). It is superficially at least very similar to *Lilium pumilum*, except that its flowers are erect and the bulb has more scales; its erect flowers may mean that it is really more closely related to the lilies of section Pseudolirium. As it appears to be so distinct from all other species of the genus, I have placed it in a new section of its own.

The *Flora RPS* does not give Hubei or Yunnan among the areas of distribution of this species. I was myself at first very dubious about the records of this lily from Yunnan, but I have now examined specimens collected by E. E. Maire near Dongchuan, and cannot dispute the identification. His notes state that it was a rare or very rare plant there. I have also examined some of E. H. Wilson's collections of *Lilium concolor* from western Hubei, where it would appear to grow much more robustly than it does where I have seen it wild in Shandong. Wilson says that it is also found in Hunan, and probably in Zhejiang. I have not seen any specimens from Hunan, and only one from Zhejiang (Ningbo). It is a distinct possibility that such specimens may have been collected from Chinese gardens rather than from the wild. This lily has been cultivated for ornament by the Chinese for at least a thousand years.

Lilium concolor was first introduced to cultivation in Britain in about 1805, and has persisted in western gardens, though now being rather rarely seen. Forms of it were awarded an FCC in 1896 and an AM in 1927. It tends not to be long-lived in cultivation, so seed should be sown every year if possible in order to maintain stocks. It is a beautiful little lily with its brilliantly-coloured, starry flowers, and is worth a little effort and perseverance. Individual bulbs can be persuaded to grow and flower for at least a few years in a sharply-drained, sunny position. It is quite at home in the rock garden, but must be sited carefully because of its vivid coloration.

II. Section Pseudolirium Endlicher

2 (11). **Lilium dauricum** Ker-Gawl. in *Bot. Mag.* sub t. 1210. 1809; et in *Bot. Reg.* 7: sub t. 594. 1821; Woodc. et Stearn, *Lil. World* 205, f. 40. 1950.
– *L. pensylvanicum* Ker-Gawl. in *Bot. Mag.* t. 872. 1805.

Bulb ovoid-globose, *c.* 1.5cm tall. *c.* 2cm in diameter; scales broadly lanceolate, 1–1.4cm long, 5–6mm wide, white, jointed or sometimes unjointed. Stem 50–70cm high, ribbed. Leaves scattered, the 4–5 leaves at the top of the stem in a whorl, with a cluster of white woolly hairs at the base of each leaf, margins papillose, sometimes also sparsely white-lanate. Bracts leaf-life, 4cm long; pedicels (1–) 2.5–8.5cm long, white-lanate; flowers terminal, 1–2, vermilion or red, with purple-red spots; outer whorl perianth segments oblanceolate, acuminate at the apex, angustate at the base, 7–9cm long, 1.5–2.3cm wide, exterior surface white-lanate; inner whorl perianth segments slightly narrower, nectaries with deep purple papillae on each side; stamens converging towards the centre, filaments 5–5.5cm long, glabrous, anthers *c.* 1cm long; ovary cylindric, *c.* 1.8cm long, 2–3mm broad; style 2 or more times longer than the ovary, stigma swollen, trilobed. Capsule oblong, *c.* 4–5.5cm long, 3cm broad.

Flowering period: June–July. Fruiting period: August–September.

Occurs in Heilongjiang, Jilin, Liaoning, Inner Mongolia and Hebei. It grows among shrubs on mountain slopes, in open woodland, by the sides of roads and in damp meadows, at 450–1500 metres above sea level. It is also found in Korea, Japan, Mongolia and the USSR. The bulb contains starch and can be eaten or used to make wine, or as medicine.

† This lily has been much used in producing hybrids, but is itself rarely seen now in cultivation. Given optimum conditions, it may be much more robust than the above description (based on wild specimens) suggests, and will commonly bear three or four flowers to a stem, or occasionally even as many as a dozen. It is not a difficult lily to grow, tends to have a stoloniform stem which produces offset bulblets quite plentifully, and may flower in just two years from seed. If in addition it is considered that this was probably the first East Asian lily to reach western gardens, at least as early as 1743 (perhaps from Siberian sources), it is curious that it is now so little grown. Undoubtedly it has been supplanted in popularity by its hybrids, which are

so numerous and common. Yet it is an attractive lily, and deserves to be restored to favour.

III. Section Regalia Baranova.

3. (1). **Lilium brownii** F. E. Brown ex Miellez in *Cat. Expos. S. Hort. Lille* 1841; Spae in *Ann. S. Agr. Gand.* 1: 437, t. 41. 1845. – *L. australe* Stapf in *Gard. Chron.* ser. 3, 70: 101. 1921. – *L. brownii* var. *australe* (Stapf) Stearn in Woodc. et Stearn, *Lil. World* 165. 1950.

Bulb globose, 2–4.5cm in diameter; scales lanceolate, 1.8–4cm long by 0.8–1.4cm wide, unjointed, white. Stem 0.7–2m tall, sometimes streaked with purple, sometimes papillose. Leaves scattered, frequently decreasing in size gradually from bottom to top of the stem, lanceolate or narrowly lanceolate to linear, 7–15cm long, (0.6–) 1–2cm wide, tip acuminate, base angustate, 5–7-veined, entire, glabrous on both surfaces. Flowers solitary or several sub-umbelliformly disposed; pedicels 3–10cm long, slightly curved; bracts lanceolate, 3–9cm long, 0.6–1.8cm wide; perianth trumpet-shaped, fragrant, milk-white, suffused with purple on the exterior, unspotted; perianth segments turned outwards or recurved at the tips, but not revolute, 13–18cm long, those of the outer whorl 2–4.3cm wide, acute at the tip, those of the inner whorl 3.4–5cm wide, papillose on each side of the nectary; stamens curved upwards, filaments 10–13cm long, densely pilose from the middle down, or occasionally sparsely pilose or glabrous; anthers long-ellipsoid, 1.1–1.6cm long; ovary cylindric, 3.2–3.6cm long, 4mm broad, style 8.5–11cm long, stigma trilobed. Capsule oblong, 4.5–6cm long, *c.* 3.5cm broad, ribbed, many-seeded.

Flowering period: May–June. Fruiting period: September–October.

Occurs in Guangdong, Guangxi, Hunan, Hubei, Jiangxi, Anhui, Fujian, Zhejiang, Sichuan, Yunnan, Guizhou, Shaanxi, Gansu and Henan. It grows on mountain slopes, in thickets, by roadsides, by streams or in rock crevices, at an altitude of (100–) 600–2150 metres above sea level. The bulb has a high starch content and can be eaten, and is also used as medicine.

Fig. 3.
1–6 *Lilium brownii* F. E. Brown ex Miellez var. *viridulum* Baker: 1 Upper part of plant; 2 Bulb; 3 Pistil; 4 Stamen; 5 Inner perianth segment; 6 Outer perianth segment.
7 *L. brownii* F. E. Brown ex Miellez var. *brownii*: Leaf.

L*ilium dauricum* Ker-Gawl.
a group of several bulbs flowering in the Royal Botanic Garden, Edinburgh.

Lilium brownii F. E. Brown ex Miellez (above)
flowering on Kwun Yam Shan in the New Territories of Hong Kong.

Lilium formosanum Wallace (right)
the small, hardy form of this lily flowering in cultivation in the south of England.

Lilium lancifolium Thunb. (left)
flowering in the Botanic Garden at Lu Shan, Jiangxi province.

Lilium pumilum Delile (above)
a strong specimen with five flowers growing among bushes in the hills near the Sleeping Buddha Temple, west of the city of Beijing.

N*omocharis aperta* (Franch.) Wilson (left)
this is one of the easier species of its genus to cultivate. It is seen here growing in the peat beds of the Royal Botanic Garden, Edinburgh.

N*omocharis pardanthina* Franch. (above)
this beautiful lily has for long been grown in western gardens as N. *mairei* Levl. Here it is flowering in the peat beds at the Royal Botanic Garden, Edinburgh, from a collection made in the Cang Shan range above Dali, Yunnan province, in 1981 (SBEC 503).

Notholirion bulbuliferum (Lingelsh.) Stearn (left)
this plant, flowering in the peat beds at the Royal Botanic Garden, Edinburgh, was collected in the Cang Shan, Yunnan province, in 1981, by the joint Sino-British Expedition (SBEC 963).

Notholirion campanulatum Cotton et Stearn (above)
this was also collected in the Cang Shan in 1981, and was photographed at the Royal Botanic Garden, Edinburgh (SBEC 571).

1
2
3
4
5
6
7

var. **viridulum** Baker in *Gard. Chron.* ser. 2, 24: 134. 1885; et ser. 3, 10: 225. 1891. – *L. brownii* var. *colchesteri* (Van Houtte) Wilson ex Elwes in *Gard. Chron.* ser. 3, 70: 101. 1921, in obs.

This variety differs from the type in having oblanceolate to obovate leaves.

Occurs in Hebei, Shanxi, Henan, Shaanxi, Hubei, Hunan, Jiangxi, Anhui and Zhejiang. It grows on grassy mountain slopes, in open woodland, on the sides of mountain valleys, at the edges of fields or near villages, and is also cultivated, at 300–920 metres above sea level. Fresh flowers contain essential oils, which can be used to make perfume. The bulbs have a high starch content, and are a famous delicacy; they are also used as medicine, effective in moistening the lungs and stopping coughing, and as an antipyretic, sedative, diuretic, etc.

† Both the varieties of *Lilium brownii* are extremely widespread in China, and are still quite common in some areas, despite having been dug for centuries for their edible bulbs. I have seen var. *viridulum* flowering at the end of July on the Lu Shan range in Jiangxi province, where it grew on steep rocky slopes at about a thousand metres altitude among dense scrub. The stems were almost two metres tall, and carried the flowers well clear of the tangle of bushes and vines through which they grew. There were also some plants of this lily growing at the edges of fields of tea bushes cultivated nearby. Most of the stems bore two or three flowers, though some had four or five. The flowers were generally of a dusky shade of reddish-purple on the exterior, but with considerable variation in the quantity of this coloration. Inside they were buff-yellow when first open, soon becoming creamy-white. I have also seen plants of the variety formerly called var. *australe* in flower on the hillsides of the New Territories in Hong Kong a little earlier in July. The stems of these plants were shorter (in contradiction to what is stated of this variety in cultivation), only about one and a half metres high, carrying similar numbers of flowers. These flowers were much whiter, tinged with a pinkish shade on the outside, and with no hint of yellow within. The lilies were growing on quite steep slopes, amongst sparse scrub and dense growth of *Dicranopteris* fern.

The *Flora RPS* merges var. *australe* with the type variety. It is very probable that there is continuous variation which renders this necessary. The description (in 1841) of typical *Lilium brownii* was in fact based on a plant of rather uncertain origin, which has even been suspected of being a hybrid, between var. *viridulum* and *Lilium formosanum*. Such hybridity is most unlikely, as *Lilium formosanum* was little known on the mainland until recent times. The self-sterility of the original introduction of var. *brownii* was almost certainly due to its being a cultivated clone, long propagated by vegetative means. Different wild forms of the species may well have been involved in its parentage.

This lily has been in cultivation in China for more than a millennium, and it is very likely that both its varieties came into cultivation in the west from Chinese nursery gardens. The type variety arrived in England in about 1837. It had been preceded by var. *viridulum*, which arrived in 1804 aboard an East Indiaman from Canton. Both varieties are supremely beautiful plants, with flowers that add strong fragrance to their other fine qualities. A form shown in 1895 was awarded an FCC. Unfortunately, they have proved difficult to grow in Britain, though propagated in large quantities by Dutch and Belgian nurserymen. They require the sharpest possible drainage.

4 (2). **Lilium regale** Wilson in *Horticulture* 16: 110. 1912, in nota; et Wilson, *Lil. East. As.* 37, t. 5. 1925. – *L. myriophyllum* Wilson in *Flora et Silva* 3: 330, t. 1. 1905.

Bulb broadly ovoid, *c.* 5cm tall, 3.5cm in diameter; scales lanceolate, 4–5cm long, 1–1.5cm wide. Stem *c.*50cm tall, papillose. Leaves scattered, numerous, narrowly linear, 6–8cm long, 2–3mm wide, one-nerved, margins and underside of midrib papillose. Flowers solitary to several, very fragrant when open, trumpet-shaped, white with a yellow throat; outer perianth segments lanceolate, 9–11cm long, 1.5–2cm wide; inner perianth segments obovate, acute at the apex, angustate at the base, nectaries without papillae on each side; filaments 6–7.5cm long, scarcely papillose, anthers ellipsoid, 0.9–1.2cm long, *c.* 3mm broad; ovary cylindric, *c.* 2.2cm long, *c.* 3mm broad; style 6cm long, stigma swollen, 6mm broad.

Flowering period June–July.

Occurs in Sichuan. It grows on mountain slopes among rocks and on river-banks, at 800–2,500 metres above sea level.

† Since its introduction to Europe and America by E. H. Wilson in 1905 to 1910, this has proved to be one of the most amenable of Chinese lilies. It will grow in a range of situations, in any reasonably well-drained soil and will often flower reliably for many years with a minimum of attention. Its beauty is also undeniable, and despite the modern fashion for hybrids it has held its place in gardens quite well. Wilson wrote that he 'would proudly rest his reputation with the Regal Lily', and this high regard has been fully justified. The only problem commonly encountered in growing *Lilium regale* is that its young shoots are easily damaged by late spring frosts, from which they should be protected. It is otherwise entirely hardy, and will stand any amount of winter freezing.

Its natural range of distribution is strangely limited for so adaptable a plant. It occurs over a 50-mile stretch of the valley of the River Min in the vicinity of Wenchuan (on the road from Guanxian to Songpan) in western Sichuan, and apparently nowhere else.

5 (3). **Lilium formosanum** Wallace in *Garden* 40: 442. 1891; Stapf in *Bot. Mag.* t. 9205. 1930; Woodc. et Stearn, *Lil. World* 219, f. 47. 1950. – *L. formosanum* var. *pricei* Staker in *Lily Year Book* 4: 16. 1935; Woodc. et Stearn, *Lil. World* 219, f. 46. 1950.

Bulb subglobose, 2–3cm tall, 2–4cm in diameter; scales oblong-lanceolate to lanceolate-ovate, white or tinged with yellow. Stem 20–55cm tall, sometimes tinged purple-red, sometimes papillose. Leaves scattered, linear to narrowly lanceolate, 10–12cm long, 4–7mm wide, entire, both surfaces glabrous. Flowers 1–2, sometimes 3–10 more or less umbelliformly disposed, fragrant, horizontal, trumpet-shaped, with a slender tube gradually expanding towards its upper section, white, tinged on the outside with purple-red; perianth segments revolute at the tip, 11.5–14.5cm long; outer segments oblanceolate, *c.* 2.2cm wide; inner segments spathulate, up to 3cm wide, nectarial furrow green, not or rarely indistinctly papillose; filaments *c.* 10cm long, flattened, with minute protuberances near the base, anthers oblong, *c.* 1cm long; ovary cylindric, *c.* 5cm long, 4mm broad; style 6.5cm long, glabrous, stigma swollen, trilobed. Capsule erect, cylindric, 7–9cm long, 2cm broad.

According to Wilson, the flowers open at any time of the year at low altitudes, but during July and August at 2000–3300 metres.

Occurs in Taiwan, growing at 3500 metres above sea level and below, on sunny, grassy slopes.

† The tall, lowland forms of this lily are rarely successful outdoors in Britain, probably because they flower very late (in August to September) and have no time afterwards to cease growth naturally before being cut down by frost. The dwarf, high-altitude forms, however, flower about a month earlier and are not difficult to grow, though usually short-lived. Fortunately, this lily is easily reproduced from seed, and may even flower within the first season of growth. The dwarf forms have commonly been known as var. *pricei*, and received an AM under this name in 1929. Their flowers are no smaller than those of the taller forms, and they can look very fine when correctly sited on the rock garden, in porous soil with leaf-mould incorporated. There are no differences between the forms other than the height of the stem, which is a continuously variable characteristic. It is no doubt for this reason that the *Flora RPS* does not recognize the variety.

This species has been in cultivation in Britain since 1880, and in the USA since 1918. There have been several separate introductions. In South Africa and Australia this lily has acclimatized so well that it has become naturalized in a number of areas.

6 (4). **Lilium longiflorum** Thunb. in *Trans. Linn. Soc.* 2: 333. 1794. – *L. longiflorum* var. *scabrum* Masumune in *Trans. Nat. Soc. Form.* 26: 218. 1936.

Bulb globose or subglobose, 2.5–5cm tall; scales white. Stem 45–90cm tall, green, light red at the base. Leaves scattered, lanceolate or oblong-lanceolate, 8–15cm long, 1–1.8cm wide, tip acuminate, margins entire, glabrous on both surfaces. Flowers solitary or 2–3; pedicels 3cm long; bracts lanceolate to ovate-lanceolate, *c.* 8cm long, 1–1.4cm wide; perianth trumpet-shaped, white, the outside of the tube slightly tinged with green, up to 19cm long; outer segments 2.5–4cm wide towards the tip; inner segments slightly wider than the outer ones, nectaries without papillae on each side. Filaments 15cm long, glabrous; ovary cylindric, 4cm long.

Flowering period: June–July. Fruiting period: August–September.

Occurs in Taiwan. Its distribution range also includes Japan (Ryukyu Islands).

This is a beautiful flower, and is extremely fragrant, containing essential oils which may be used for making perfume.

† *L. longiflorum* is the florists' Easter Lily, easily forced and widely grown for the cut-flower trade. Numerous cultivars have been developed and named. It is not hardy in cool climates, and needs the protection of a greenhouse, but can be successfully grown in large pots and may be brought into flower at any time of year. In cultivation it is very susceptible to virus diseases, so should be regularly raised from seed to maintain clean stocks.

It is a debatable point whether this lily is a native of Taiwan. It has been suggested by several authorities that it may well have been spread from its original island home in the Ryukyu archipelago by the agency of man, who planted it on graves, and it could certainly have reached Taiwan in this manner. Whether introduced or not, it undoubtedly grows wild on Taiwan now, though only in a restricted area in the north of the island. It came into cultivation in Japan before 1681, and probably reached mainland Chinese gardens from there, being known in China by 1688. It commonly grows on limestone soils in nature, and shows no aversity to lime in cultivation.

7 (5). **Lilium leucanthum** (Baker) Baker in *Journ. Roy. Hort. Soc.* 26: 337. 1901, pro parte (excl. *Bot. Mag.* t. 7722. 1900). – *L. brownii* var. *leucanthum* Baker in *Gard. Chron.* ser. 3, 16: 180. 1894.

Bulb subglobose, 3.5–4cm tall; scales lanceolate, 3.5cm long, *c.* 1cm wide, brownish-yellow or purple when dried. Stem 60–150cm tall, papillose. Leaves scattered, lanceolate, 8–17cm long, 6–10mm wide, margins not papillose, axils of the leaves of the upper part of the stem not bulbiliferous. Flowers solitary or 2–4; bracts oblong-lanceolate, (4–) 5–6cm long, slightly wider than the leaves, 1.2–1.6cm wide; pedicels up to 6cm long, purple; perianth trumpet-shaped, slightly fragrant, white, pale yellow inside, dorsal

4
2
1
3
6
5

ridge and adjacent area pale greenish-yellow, 12–15cm long; outer perianth segments lanceolate, 1.6–2.8cm wide; inner perianth segments spathulate, 2.6–3.8cm wide, apex obtuse, nectaries not papillose; filaments 10–22cm long, densely hairy in the lower part, anthers ellipsoid, *c.* 1cm long, 4–5mm broad, pale yellow; style up to 10cm long, hairy at the base; stigma swollen, 8mm in diameter, trilobed.

Flowering period: June–July.

Occurs in Hubei and Sichuan. It grows among herbs in mountain gullies and by rivers, at 450–1500 metres above sea level.

var. **centifolium** (Stapf) Stearn in Woodc. et Coutts, *Lil.* 213. 1935 (June), apud Sealy in *Gard. Chron.* ser. 3, 98: 144. 1935 (Aug.). – *L. centifolium* Stapf apud Elwes in *Gard. Chron.* ser. 3, 70: 101. 1921.

This variety differs from the type in that the outer surfaces of the perianth segments are purple or tinged with brown.

Flowering period: June–July.

Occurs in Gansu (Zhouqu). It grows in mountain gullies, at 2500 metres above sea level.

† The type variety of this lily first became known in western gardens when Augustine Henry sent a box of bulbs from Yichang to Kew, which arrived in 1889. There was initially much confusion about its naming, and it was placed as a variety of both *L. longiflorum* and *L. brownii* before finally being recognized as a distinct species. E. H. Wilson was familiar with it as a wild plant, saying that he had found it to be common in the mountains of western Hubei and eastern and central Sichuan to the western limits of the Red Basin. This brings its area of distribution very close to those of both *L. regale* and *L. sargentiae*, which seem to be its nearest relatives.

Lilium leucanthum was first collected in southern Gansu in 1894 by a Russian expedition, and was introduced to the Botanic Garden at Petrograd in 1895. Seeds from these plants were distributed to Germany and bulbs raised there were sent on to England in 1912 or 1913. One bulb flowered with A. Grove in 1914 and was described as 'a magnificent lily'. This must have been var. *centifolium*, though at the time it was grown under the name of '*L. brownii kansuense*'. It does not seem to have persisted, and the stocks of var. *centifolium* later grown in Britain probably all derived from Farrer's collection made in 1914 from a cottage garden near Siku (Zhouqu), southern Gansu. Though Farrer never saw this lily growing wild, and doubted its

Fig. 4.
1–4 *Lilium leucanthum* Baker: 1 Upper part of plant; 2 Leaf; 3 Bract; 4 Stamen and pistil.
5–6 *L. regale* Wilson: 5 Upper part of plant; 6 Leaf.

existence as a wild plant in the area where he saw it cultivated, both the Russian collections and Chinese records indicate that it does, in fact, occur naturally in this region. Zhouqu is situated only twenty miles or so from the Gansu-Sichuan border, and is therefore not very remote from the recorded area of distribution of the type variety.

Wilson sent several hundred bulbs of *L. leucanthum* to the Arnold Arboretum in 1908 and 1910, but it apparently was not hardy in the Boston area and most of the bulbs were killed by winter frosts. It is, however, hardy at least in the southern half of Britain, though young bulbs should be grown on in pots until large enough to be planted deeply (about 13cm down). When once established, this lily may be quite long-lived, and has been known to grow and flower for more than a dozen years.

It has been extensively used in hybridization, both in the USA and Britain. Probably most of the popular trumpet hybrids now in cultivation, such as 'Black Dragon' and the Olympic Hybrids, have this lily in their parentage.

8 (6). **Lilium sulphureum** Baker apud Hook. f., *Fl. Brit. Ind.* 6: 351. 1892 (Jul.); Baker in Bot. Mag. t. 7257. 1892 (Oct.). – *L. myriophyllum* Franch. in *Journ. de Bot.* 6: 313. 1892 (Sept.).

Bulb globose, 3–5cm tall, 5.5cm in diameter; scales ovate-lanceolate or lanceolate, 2.5–5cm long, 0.8–1.6cm wide. Stem 80–120cm tall, papillose. Leaves scattered, lanceolate, 7–13cm long, 1.3–1.8 (–3.2)cm wide, with bulbils in the axils of the upper leaves. Bracts ovate-lanceolate or elliptic; pedicels 4.5–6.5cm long; flowers usually 2, trumpet-shaped, fragrant, white, yellow within the tube; perianth segments 17–19cm long; outer segments oblong-oblanceolate, 1.8–2.2cm wide; inner segments spathulate, 3.2–4cm wide, nectaries not papillose on each side; filaments 13–15cm long, glabrous or occasionally sparsely hairy; anthers long-oblong, *c.* 2cm long; ovary cylindric, 4–4.5cm long, 2–5mm broad, purple; style 11–12cm long, stigma swollen, *c.* 1cm in diameter.

Flowering period: June–July.

Occurs in Yunnan, Guizhou, Sichuan and Guangxi. Grows by the sides of roads, on grassy slopes, or on mountain slopes in shady places among open woodland, at 90–1890 metres above sea level. It is also found in Burma. The bulbs can be used medicinally.

This species is very close to *L. sargentiae* Wilson, but differs in having

Fig. 5.
1–4 *Lilium sulphureum* Baker: 1 Upper part of plant; 2 Outer perianth segment; 3 Inner perianth segment; 4 Stamen and pistil.
5–6 *L. sargentiae* Wilson: 5 Leaf and bulbil; 6 Inner perianth segment and stamen.

1
2
3
4
5
6

glabrous filaments and slightly larger flowers, with perianth segments 17–19cm long.

† This is a very beautiful lily, but is not very hardy. In the wild it does not appear to occur north of about 26° 30′, and never at altitudes great enough for it to be exposed to very severe freezing. Indeed, in much of its range it is only rarely exposed to frost and often does not die down at the end of the season until well into November. In cultivation in cooler climates, it is undoubtedly more easily managed in large pots in a greenhouse. It has, however, been successfully grown in the open garden in southern Britain on a number of occasions. It usually flowers rather late in cultivation, in late August or even September, and only rarely sets any seed. This has meant that it has been propagated mainly by bulbils, with a consequently high risk of build-up of disease, and as a result is now very rare in British gardens.

9 (7). **Lilium sargentiae** Wilson in *Gard. Chron.* ser. 3, 51 : 385. 1912; et in *Horticulture* 18: 169. 1913; Woodc. et Stearn, *Lil. World* 331, f. 104. 1950. – *L. formosum* Franch. in *Journ. de Bot.* 6: 317. 1892.

Bulb subglobose or broadly ovoid, 4–4.5cm tall, 5–6cm in diameter; scales lanceolate, 3.5–4cm long, 1.5–1.7cm wide. Stem 45–160cm tall, papillose. Leaves scattered, lanceolate or oblong-lanceolate, 5.5–12cm long, 1–3cm wide, with bulbils in the axils of the upper leaves. Bracts ovate-lanceolate, 5–6cm long, 1.2–2cm wide; pedicels 5.5–8.5cm long; flowers 1–4, trumpet-shaped, white, pale green towards the base, slightly revolute at the apex; outer perianth segments oblanceolate, 14–16 (–17)cm long, 2–2.8cm wide; inner perianth segments wider than the outer segments, narrowly obovate-spathulate, nectaries yellowish-green, not papillose; filaments 11–13cm long, densely hairy in the lower part; anthers oblong, 1.4–2cm long, pollen brownish-yellow; ovary cylindric, 3.5–4.5cm long, 3–5mm in diameter, purple; style 10–11cm long, slightly curved at the upper end, stigma swollen, 8–10mm in diameter, trilobed. Capsule oblong, 6–7cm long, *c.* 3.5cm broad.

Flowering period: July–August. Fruiting period: October.

Occurs in Sichuan. Grows among herbs on mountain slopes, or on the edges of thickets, at 500–2000 metres above sea level.

This species is very close to *L. sulphureum* Baker, but differs in having filaments which are densely hairy in the lower part, and perianth segments 14–17cm long.

† *Lilium sargentiae* is a very fine trumpet-lily, and one of the easier members of its section to cultivate. It dislikes a limy soil, and will not tolerate excessive wetness in winter, but is otherwise not too hard to please. Unfortunately, it only sets seed in Britain on the rare occasions when the summer is sufficiently hot. Being bulbiliferous, it is not hard to propagate, but seed

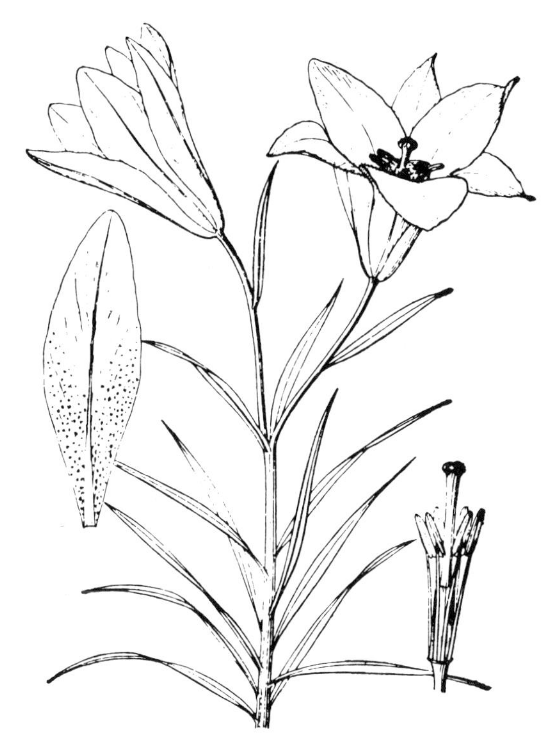

Fig. 6.
Lilium bakerianum Coll. et Hemsl.

should be sown whenever available to limit the spread of virus disease.

Its area of natural distribution is among the mountains of western Sichuan, from the vicinity of Guanxian in the north to Emei Shan and Wa Shan in the south, extending from the edge of the Sichuan Basin westwards to Luding on the Dadu River. It was E. H. Wilson who first collected it, and introduced it in quantity into both Britain and the USA. It was at first confused with *L. leucanthum*, but was described as a new species in 1912. It was awarded an FCC (as *L. leucanthum*) when shown by Messrs. Veitch in 1905.

It has been important in the breeding of hybrid lilies, especially through its cross with *L. henryi* which resulted in the Aurelianense hybrids.

IV. Section Lophophorum (Bur. et Franch.) Wang et Tang

10 (15). **Lilium bakerianum** Coll. et Hemsl. in *Journ. Linn. Soc. Bot.* 28: 138, t. 22. 1898.

Bulb broadly ovoid to sub-globose, 2.5–3cm tall, diameter *c.* 2.5cm; scales ovate or ovate-lanceolate, 2–2.2cm long, 7–10mm wide, white. Stem

60–90cm tall, papillose. Leaves scattered on the middle and upper parts of the stem, linear or linear-lanceolate, 4–7.5cm long, 4–7mm wide, tip acuminate, margins and underside of main vein papillose. Flowers 1–3, campanulate, sub-erect or nodding, white, with purple-red spots inside; outer whorl perianth segments lanceolate, 6.5–8.3cm long, 1.4–1.8cm wide, tip acute; inner whorl perianth segments wider, oblanceolate or oblanceolate-spathulate, 6.5–8cm long, 1.1–2.3cm wide, tip more or less rounded, nectaries not papillose on each side; filaments subulate, *c.* 3cm long, glabrous, anthers 1.6cm long, orange; ovary cylindric, 1.7–2cm long, 2–4mm broad; style 2.2–2.6cm long, stigma sub-globose, 2.5–5mm in diameter, trilobed. Capsule oblong, *c.* 3.5cm long by *c.* 2.5cm broad.

Flowering period: July.

Occurs in Yunnan (north-west part) and Sichuan (western part). It grows at the edges of woodland, at about 2800 metres above sea level. It is also distributed in northern Burma.

var. **aureum** Grove et Cotton in *Lily Year Book* 8: 127. 1939.

This variety differs from the type in having pale yellow flowers spotted with purple inside.

Occurs in Yunnan (north-west part) and Sichuan (south-west part). It grows on grassy slopes among woodland or at the edges of thickets, at 2000–2420 metres above sea level.

var. **delavayi** (Franch.) Wilson, *Lil. East. As.* 43. 1925. – *Lilium delavayi* Franch. in *Journ. de Bot.* 6: 314. 1892.

This variety differs from the type in having yellowish-green or olive-green to pale green flowers with red-purple or bright red spots inside.

Occurs in Yunnan, Sichuan and Guizhou. It grows among woodland on mountain slopes, or on grassy slopes, at 2500–3800 metres above sea level.

var. **rubrum** Stearn in *Gard. Chron.* ser. 3, 124: 4. 1928.

This variety differs from the type in having red or pink flowers, with red or purple-red spots.

Occurs in Yunnan and Guizhou. It grows at the edges of mixed woodland, on the banks of streams or in grassland on mountain slopes, at 1500–2000 metres above sea level.

var. **yunnanense** (Franch.) Sealy ex Woodc. et Stearn, *Lil. World* 151. 1950. – *Lilium yunnanense* Franch. in *Journ. de Bot.* 6: 314. 1892.

This variety differs from the type in having unspotted, white or pale rose-pink flowers, and leaves white-pilose on both surfaces and with papillose margins.

Occurs in Yunnan (north-west part) and Sichuan (south-west part). It grows in pine woods or in meadows, at 2000–2800 metres above sea level.

† This is a widely-distributed and very variable lily, with a confused taxonomic history. It was first discovered in the Shan Hills of Upper Burma in 1888, and various later collections from western China were at first not associated either with the Burmese plant or, very often, with each other. Several different specific names thus came to be applied to this lily. During the first half of this century, however, it was seen and collected on numerous occasions, and the confusion was gradually resolved. The five varieties now recognized represent the extremes of variation within the species, and intermediate forms occur. There is considerable overlap in the geographical distributions of the varieties, and two or more quite commonly grow together, particularly near the central area of the overall range of distribution, in north-west Yunnan. The variety *delavayi* is perhaps the most widespread, ranging from around Muli and Yanyuan in south-west Sichuan across much of Yunnan province, and from near the Burmese border west of Tengchong eastwards to south-west Guizhou. The type variety occurs at the northern extreme of the range of the species, in the vicinity of Kangding in Sichuan, and has also been collected from the Lijiang range in north-west Yunnan and from Burma. Var. *aureum* also grows in the Lijiang range, as well as in the Dali area and the mountains west of Jianchuan in Yunnan. Var. *yunnanense* would appear to be common in north-west Yunnan, with its southern limit near Songgui (*c.* 26° 20′ N.), and extending to the Muli region of Sichuan. The variety with the most easterly range is var. *rubrum*, which occurs in eastern Yunnan from the vicinity of Dongchuan in the north to Mengzi in the south, and as far east as the Anshun area in Guizhou.

All the varieties of *Lilium bakerianum*, with the probable exception of var. *rubrum*, have been in cultivation in the west. Quite a number of introductions were in fact made, by several collectors from various different localities. But this lily proved extremely difficult to manage, and most stocks have long since dwindled away. Probably only the type variety now maintains a precarious existence in British gardens. Many of the introduced stocks proved to be rather tender, even those from fairly high altitudes at the northern edge of the lily's range. But even when carefully cossetted in the greenhouse, they rarely flourished for more than a few years. It seems that *Lilium bakerianum* is likely to remain one of the elusive treasures of the lily world.

11 (9). **Lilium nanum** Klotz. et Garcke, *Bot. Erg. Reise Pr. Waldemar* 53. 1862. – *Nomocharis nana* (Klotz. et Garcke) Wilson, *Lil. East. As.* 13. 1925.

Bulb oblong, 2–3.5cm tall, 1.5–2.3cm in diameter; scales lanceolate, 2–2.5cm long, 5–8mm wide; stem-roots absent. Stem 10–30cm tall, glabrous. Leaves scattered, linear, 6–11, 4–8.5cm long, 2–4mm wide, the 2–3 near the base

3
5
6
1
2
4
9
8
7

comparatively short and wide. Flower solitary, campanulate, nodding; perianth segments pale purple or purplish-red, spotted with deep purple inside; outer segments elliptic, 2.5–2.7cm long, 1–1.2cm wide; inner segments somewhat wider than the outer ones, nectaries with fimbriate projections on each side; stamens converging, filaments subulate, 1–1.2cm long, glabrous; anthers ellipsoid, *c.* 6mm long; ovary cylindric, 1cm long, 3–6mm wide; style 4–6mm long, stigma swollen, 3–4mm in diameter. Capsule oblong, 2.8–3.5cm long, 2–2.5cm wide, yellow, the ribs tinged with purple.

Flowering period: June. Fruiting period: September.

Occurs in Tibet (south and south-east parts), Yunnan (north-west part) and Sichuan (western part). It grows on grassy mountain slopes, in thickets or at the edges of woods, at 3500–4500 metres above sea level. It is also distributed in Nepal, Sikkim and Bhutan.

var. **flavidum** (Rendle) Sealy in *Bot. Mag.* t. 218. 1952. – *Fritillaria flavida* Rendle in *Journ. of Bot.* 44: 45. 1906. – *Lilium euxanthum* (W. W. Sm. et W. E. Evans) Sealy in *Kew Bull.* 289, f. 4. 1950. – *Nomocharis euxantha* W. W. Sm. et W. E. Evans in *Notes Bot. Gard. Edinb.* 15: 4, t. 201. 1925.

This variety differs from the type in having yellow flowers.

Occurs in Tibet (south-eastern part) and Yunnan. It grows at 3800–4280 metres above sea level at the edges of woods and in alpine meadows. It is also distributed in Burma.

var. **brevistylum** Liang, *Flora RPS* 14: 283. 1980.

This variety differs from the type in having yellow flowers slightly tinged with purple, a short style, *c.* 1mm long, filaments shorter than the ovary, and bulb-scales tinged with purple.

Occurs in Tibet (Zayü). It grows on the edges of woods at 4280 metres above sea level.

† This is a pretty little lily in all its forms, though by no means as showy as many of its near relatives. It has been collected many times, from China and in recent years from the Nepal and Sikkim Himalaya, and is quite well-established in cultivation. It is most likely to succeed if planted among dwarf rhododendrons in a raised peat bed, and in Britain undoubtedly prefers the

Fig. 7.
1–3 *Lilium nanum* Klotz. et Garcke: 1 Whole plant; 2 Inner perianth segment; 3 Outer perianth segment.
4–9 *L. lophophorum* (Bur. et Franch.) Franch.: 4 Whole plant; 5 Outer perianth segment; 6 Inner perianth segment; 7–9 Leaves.

cool climate of the north. Seed is often set and usually germinates freely, but the growth of seedlings is usually rather slow, and they often take four or five years to reach flowering size.

Lilium nanum is very widespread in the wild, occurring along the Himalaya from Himachal Pradesh in the west to an eastern limit at about 101° E. It is recorded from the Lijiang range in Yunnan, and may also occur as far to the north-east as the vicinity of Songpan in Sichuan. The pale-flowered var. *flavidum* is often found with or near the type variety, but does not extend so far to the west, its limit being in eastern Nepal. It is fairly common in Sikkim, south-east Tibet and north-west Yunnan.

12 (8). **Lilium lophophorum** (Bur. et Franch.) Franch. in *Journ. de Bot.* 12: 221. 1898; Wilson, *Lil. East. As.* 104, t. 16. 1925. – *Fritillaria lophophora* Bur. et Franch. in *Journ. de Bot.* 5: 153. 1891. – *Nomocharis lophophora* (Bur. et Franch.) W. E. Evans in *Notes Bot. Gard. Edinb.* 15: 11–14. 1925; Woodc. et Stearn, *Lil. World* 387, f. 131. 1950. – *N. wardii* Balf. f. in *Trans. Bot. Soc. Edinb.* 27: 297. 1918. – *N. lophophora* var. *wardii* (Balf. f.) W. W. Sm. et W. E. Evans in *Notes Bot. Gard. Edinb.* 14: 120. 1925. – *Lilium lophophorum* subsp. *typicum* Sealy forma *wardii* (Balf. f.) Sealy in *Kew Bull.* 294, f. 5, e. 1950. – *L. lophophorum* subsp. *typicum* forma *latifolium* Sealy in *Kew Bull.* 294, f. 5, f–k. 1950.

Bulb subovoid, 4–4.5cm tall, 1.5–3.5cm in diameter; scales rather loose, lanceolate, 3.5–4cm long, 6–7mm wide, white, stem-roots absent. Stem 10–45cm tall, glabrous. Leaves very variable, clustered to scattered, lanceolate, oblong-lanceolate or long-lanceolate, 5–12cm long, 0.3–2cm wide, apex obtuse, acute or acuminate, base angustate, margins papillose, 3–5-nerved. Flowers usually solitary, occasionally 2–3, nodding; pedicels 9–15cm long; bracts leaf-like, lanceolate, 5–13cm long, 3–10mm wide; perianth yellow, pale yellow or pale yellowish-green, with extremely sparse purple-red spots or unspotted; perianth segments lanceolate or narrowly ovate-lanceolate, 4.5–5.7cm long, 0.9–1.6cm wide, apex long-acuminate, inner perianth segments with fimbriate projections on each side of the nectaries; stamens converging, 1.5–2cm long, filaments subulate, glabrous, anthers ellipsoid, 0.7–1cm long; ovary cylindric, 1–1.2cm long, 3–4mm broad; style *c.* 1cm long, stigma swollen, capitate. Capsule oblong, 2–3cm long, 1.5–2cm broad, tinged purple when ripe.

Flowering period: June–July. Fruiting period: August–September.

Occurs in Sichuan, Yunnan and Tibet. It grows in alpine meadows, woodland or thickets on mountain slopes, at 2700–4250 metres above sea level.

var. **linearifolium** (Sealy) Liang, *Flora RPS* 14: 129. 1980. – *L. lophophorum* (Bur. et Franch.) Franch. subsp. *linearifolium* Sealy in *Kew Bull.* 294, f. 4, p–q. 1950.

This variety differs from the type in having altogether 15–16 linear leaves clustered together in the middle and upper parts of the stem, and in having yellow flowers with conspicuous purple spots.

† *Lilium lophophorum* is an odd little lily, which E. H. Wilson even suggested should be placed in a genus on its own. In reality, it is only its elongated petals which give it its aberrant appearance, and in all other respects it is clearly closely related to *LL. nanum* and *oxypetalum*. Like them, it has in the past been variously assigned to the genera *Fritillaria* and *Nomocharis* as well as *Lilium*. The species of section Lophophorum are undoubtedly closely related to *Nomocharis*, but I believe that their resemblance to *Fritillaria* is mainly superficial. The bulbs are certainly quite distinct, being composed of numerous scales, as is usual in *Lilium*, rather than the few very thick scales typical of the genus *Fritillaria*. Those taxonomists who placed these species in *Fritillaria* had usually only seen incomplete specimens lacking the bulb.

Lilium lophophorum has been frequently collected in western China. Wilson found it to be common in the mountains around Kangding, and it was also collected further west, between Litang and Batang. In Yunnan, many collections were made in the Lijiang mountains, and in the area northwards across the Zhongdian plateau to the Yunnan-Tibet border region. There are also collections from around Yongning and Muli. Its most southerly location seems to be on the Lancang Jiang-Jianchuan (Mekong-Chienchuan) divide at about 26° 30′ N. Wilson stated that it was also found in southern Gansu, but I have seen no specimens from there (or anywhere very near there), and think that this must be considered doubtful. The *Flora RPS* gives only Sichuan, Yunnan and Tibet.

A number of introductions of this lily to both Britain and the USA were made by Wilson, Forrest, Rock and others, but it never persisted, and has long been lost to cultivation.

13 (12). **Lilium souliei** (Franch.) Sealy in *Kew Bull.* 296, f. 3. 1950. – *Fritillaria souliei* Franch. in *Journ. de Bot.* 12: 221. 1898. – *Nomocharis souliei* (Franch.) W. W. Sm. et W. E. Evans in *Notes Bot. Gard. Edinb.* 14: 102. 1925.

Bulb more or less narrowly ovoid, 2.5–3cm tall, 1.3–1.8cm in diameter; scales lanceolate, 1.5–3cm long, 6–10mm wide, white. Stem 10–30cm tall, glabrous. Leaves scattered, 5–8, long-elliptic, lanceolate or linear, 3–6cm long by 0.6–1.5cm wide, entire or with sparsely papillose margins. Flower solitary, campanulate, nodding, purple-red, unspotted, the colour becoming paler inside towards the base of the perianth; outer whorl perianth segments elliptic, 2.5–3.5cm long, 9–12mm wide, apex acute, with a short point; inner whorl perianth segments 1–1.8cm wide, obtuse, nectaries not papillose; stamens converging, filaments 1.2–1.4cm long, glabrous, anthers 5–7mm long by 2–3mm broad, purple-black; style up to 1.2cm long, stigma slightly swollen. Capsule sub-globose, 1.5–2cm long, 1.5–2cm broad, tinged with purple.

4
3
5
2
1
6
8
7

Flowering period: June–July. Fruiting period: August–October.
Occurs in Sichuan, Yunnan and Tibet. It grows on grassy mountain slopes or at the edges of thickets, at 1200–4000 metres above sea level.

† This is yet another small alpine lily which has only transiently been in cultivation, and remains tantalizingly unobtainable. It was collected quite frequently in the early decades of this century until as late as 1947, but was never grown successfully in western gardens. It is distributed from the Muli area in Sichuan and the Lijiang range in Yunnan north-westwards into south-east Tibet, as far west as about 93° 50′ E., on the mountains south of the Yarlung Zangbo (Yalu Tsangpo) River. It has also been found on the Burma-Tibet border, and may occur in northern Burma.

14 (16). **Lilium sempervivoideum** Lévl. in *Bull. Géogr. Bot.* 25: 38. 1915; et *Cat. Pl. Yunnan.* 166, f. 39. 1917.

Bulb sub-globose, 2.5–3cm tall, 2.5–3cm in diameter; scales lanceolate, 2.5–3cm long, 0.5–1cm wide. Stem 20–30cm tall, papillose. Leaves 16–30, scattered on the middle part of the stem, linear, 2.5–5.5cm long, 2–4mm wide, 1-nerved, entire. Flower solitary, campanulate, white, with minute purple-red basal spots; outer whorl perianth segments lanceolate. 3.5–4cm long, 5–10mm wide; inner segments wider, narrowly elliptic-lanceolate, 1.2–1.5cm wide, nectaries not papillose; stamens converging, filaments 1.2–1.5cm long, glabrous, anthers long-oblong, 5.5–6.5cm long; ovary purple-black, *c.* 8mm long, 1.5–2.5mm broad; style 1.5cm long, stigma swollen, 3–4mm in diameter, trilobed.
Flowering period: June.
Occurs in Yunnan and Sichuan. It grows on mountain slopes in grassy places, at 2400–2600 metres above sea level.

† A very little-known lily, which has never been in cultivation in the west. It grows in north-east Yunnan and apparently also in adjacent parts of Sichuan, and would appear to be closely related to both *L. bakerianum* and *L. amoenum*. It is undoubtedly an attractive plant, but is unlikely to be very amenable to cultivation, even if it could be obtained.

Fig. 8.
1–5 *Lilium souliei* (Franch.) Sealy: 1 Bulb; 2 Upper part of plant; 3 outer perianth segment; 4 Inner perianth segment; 5 Stamen and pistil.
6–8 *L. henrici* Franch.: 6 Upper part of plant; 7 Outer perianth segment; 8 Inner perianth segment.

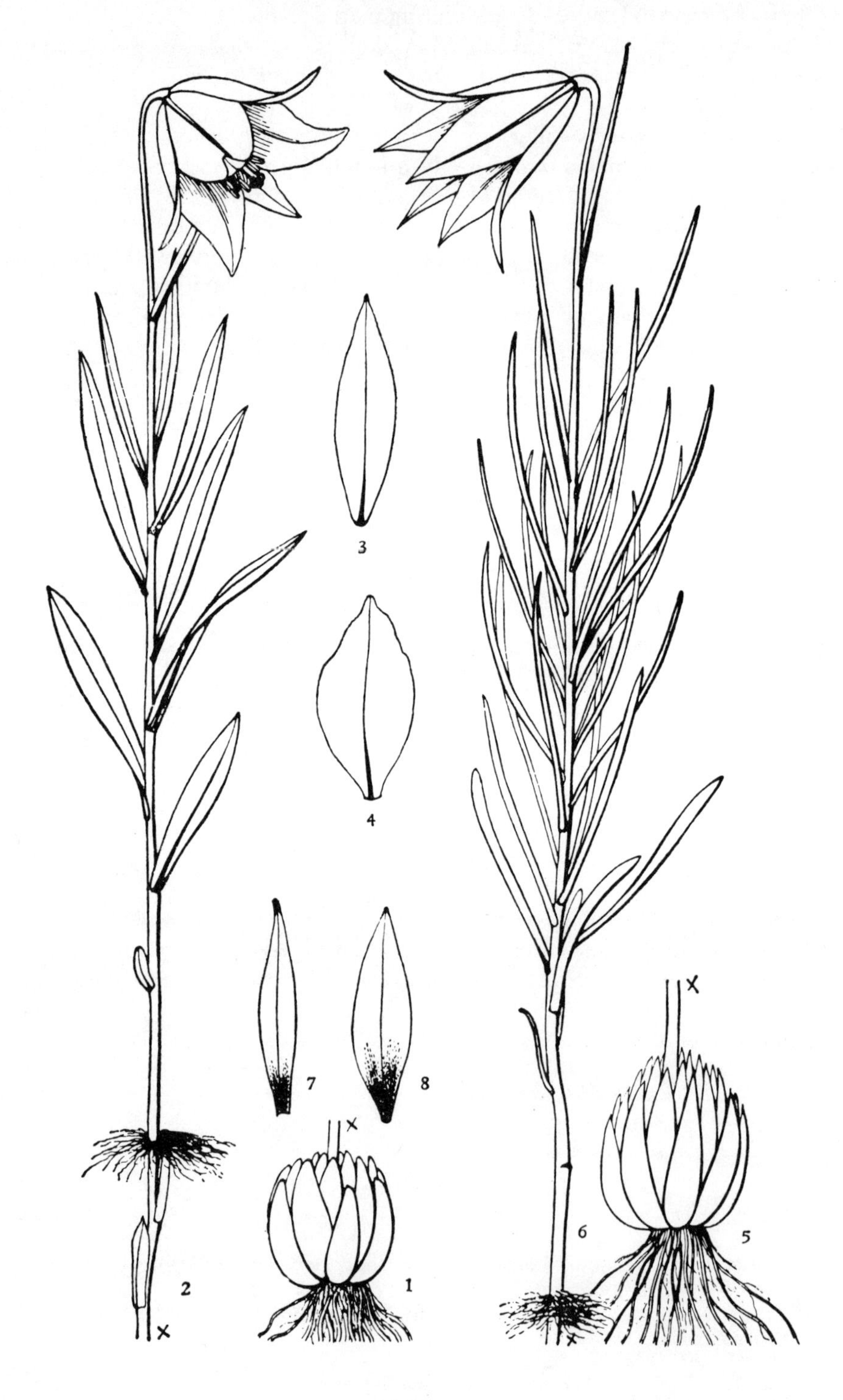
3
4
7
8
X
1
2
X
6
5
X
X

15 (17). **Lilium amoenum** Wilson ex Sealy in *Bot. Mag.* t. 93. 1949.

Bulb ovoid, 2–2.5cm tall, 2–2.2cm in diameter; scales lanceolate, 1.5–1.8cm long, 6–8mm wide, white. Stem 15–30cm tall, papillose. Leaves scattered, 8–12, long-elliptic or narrowly oblong, 2.8–4.5cm long, 2–7mm wide, glabrous, entire, with a prominent midrib. Flower solitary, fragrant, campanulate, purple-red or purple-rose, spotted with red, nodding; outer whorl perianth segments lanceolate, 3–4cm long, 9–10mm wide, slightly revolute at the apex; inner segments ovate-lanceolate or elliptic, 1.4–1.5cm wide, base angustate, apex acuminate or obtuse, nectaries green; stamens converging, filaments 1.4cm long, glabrous, anthers 5–6.5cm long; ovary cylindric, 8mm long, 2mm broad; style 1.2–1.6cm long, stigma swollen, 3mm in diameter, trilobed.

Flowering period: June.

Occurs in Yunnan. It grows in woodland, at 2100–2300 metres above sea level.

† This is another very little-known species, which was, however, introduced once into cultivation in Britain. Bulbs collected in the Lijiang area, Yunnan, flowered at Bodnant in 1938. Unfortunately they died after flowering and did not set seed. This is the only known occasion on which *Lilium amoenum* has been grown outside its native mountains.

Its area of distribution extends from Mengzi in the south-east of Yunnan province, through the mountains around Kunming to Lijiang in the north. It does not seem to be common in any part of its range.

16 (13). **Lilium paradoxum** Stearn in *Bull. Brit. Mus. Bot.* 2: 78. 1956.

Bulb small, 1.5–2.5cm tall, 1–2.5cm in diameter; scales ovate, up to 2.5cm long, 8mm wide. Stem 20–45cm tall, papillose. Leaves in whorls on the middle and upper parts of the stem, sometimes also with a few scattered leaves, obovate-lanceolate or elliptic, 4.5–5.5cm long, 1.8–2cm wide, apex acute, base sub-angustate, 5–7-nerved, glabrous. Flower solitary; pedicel up to 5cm long; perianth campanulate, patulous, purple, unspotted; perianth segments narrowly elliptic or rarely narrowly ovate, 2.5–3.5cm long, 1–1.4cm wide, nectaries not papillose; filaments *c.* 1.6cm long, glabrous, anthers linear, 6–8mm long; ovary cylindric, 6–8mm long, purple; style 1.7cm long, glabrous, stigma capitate, *c.* 6mm in diameter.

Flowering period: July.

Fig. 9.
1–4 *Lilium amoenum* Wilson ex Sealy: 1–2 Whole plant; 3 Outer perianth segment; 4 Inner perianth segment.
5–8 *L. sempervivoideum* Lévl.: 5–6 Whole plant; 7 Outer perianth segment; 8 Inner perianth segment.

Occurs in south-east Tibet. It grows in grassy places and among bushes on mountain slopes, and on rocky slopes, at 3200–3900 metres above sea level.

† *Lilium paradoxum* is a peculiar lily, for which the specific epithet is very apt. It is the only member of its section to have whorled leaves, which are in addition very much broader than in any of its near relatives. A parallel development of verticillate leaf arrangement also occurs in the genus *Nomocharis*, however. The form of flower clearly places *Lilium paradoxum* in this section.

The type specimen was collected by Ludlow, Sheriff and Elliot in the vicinity of Bomi (Pome) near the great bend of the Yarlung Zangbo in south-east Tibet. This species has never been in cultivation.

17 (14). **Lilium henrici** Franch. in *Journ. de Bot.* 12: 220. 1898; Woodc. et Stearn, *Lil. World* 225, f. 59. 1950. – *Nomocharis henrici* (Franch.) Wilson, *Lil. East. As.* 13. 1925, pro parte.

Bulb ovoid or sub-globose, 3.5cm tall, 4cm in diameter; scales lanceolate, 2.5–4cm long, 8–15mm wide. Stem 60–120cm tall, glabrous. Leaves scattered, long-lanceolate, 12–15cm long, 9–14mm wide, tip long-acuminate, 3-nerved, glabrous. Flowers more or less campanulate, except in the most exceptional cases, and usually 5–6 in a raceme; pedicels 3.5–6cm long; perianth openly campanulate, white, with an obvious deep purple-red blotch inside at the base; segments more or less oblong-lanceolate, 3.5–5cm long, 1.2–1.4 (–2)cm wide, nectaries green, not papillose; stamens converging, filaments 2cm long, glabrous, anthers *c.* 1cm long; ovary cylindric, 9–13mm long, 2–3mm broad; style 1.5–2.2cm long, stigma swollen, trilobed.

Flowering period: July.

Occurs in Yunnan (north-west part) and Sichuan (western part). It grows in mixed woodland at around 2800 metres above sea level.

var. **maculatum** (W. E. Evans) Woodc. et Stearn, *Lil. World* 226. 1950. – *Nomocharis henrici* forma *maculata* W. E. Evans in *Notes Bot. Gard. Edinb.* 15: 194. 1926.

This variety differs from the type in having a few purple-red spots as well as a large purple-red blotch on each inner perianth segment; the outer segments have blotches only.

Occurs in north-west Yunnan.

† *Lilium henrici* is the closest of all the *Lilium* species to the genus *Nomocharis*. The dividing line between the genera is, indeed, a rather artificial one. The basal blotches on the perianth segments of this species have not, however, developed into the swellings characteristic of the *Nomocharis* species, so that it is possible to separate the genera by this

feature. *L. henrici* undoubtedly represents more or less the final stage in the evolution of *Lilium* towards *Nomocharis*.

This species occurs in north-west Yunnan between about 28° 30′ N., on the Nu Jiang-Drung Jiang (Salween-Kiuchiang) divide, and about 25° 20′ N., on the Longchuan-Nu Jiang (Shweli-Salween) divide. It has been collected on the Yunnan-Burma border at Chimili, north of Pianma (Hpimaw), and may also occur across the border in Burma. The most easterly location from which I have seen a specimen is the Lijiang area, but the *Flora RPS* states that it extends into Sichuan. The variety *maculatum* is the normal form in the west of the range, on the mountains west of the Nu Jiang, and occurs with the type variety on the mountains between the Nu Jiang and Lancang Jiang (Mekong). It is not found further east.

The type variety of this lily has been grown since its introduction in the 1930s in a very few gardens in central Scotland. There seems to have been very little success with it anywhere else, either in Britain or the USA, despite the distribution of considerable quantities of seed from the Scottish stocks. It is an extremely beautiful lily, and it is to be hoped that it will continue in cultivation and perhaps eventually spread to a greater number of gardens, if the problems involved in persuading it to flourish can ever be overcome.

V. Section Archelirion Baker

18 (22). **Lilium speciosum** Thunb. in *Trans. Linn. Soc.* 2: 332. 1794.

Occurs in Japan, but not in China.

var. **gloriosoides** Baker in *Gard. Chron.* n. ser. 14: 198. 1880. – *Lilium kana-hirai* Hay., *Ic. Pl. Formos.* 2: 146. 1912. – *L. konishii* Hay. in *Journ. Coll. Sci. Tokyo* 30 (1): 364. 1911.

Bulb more or less flattened-globose, 2cm tall, 5cm in diameter; scales broadly lanceolate, 2cm long, 1.2cm wide, white. Stem 60–120cm tall, glabrous. Leaves scattered, broadly lanceolate, oblong-lanceolate or ovate-lanceolate, 2.5–10cm long, 2.5–4cm wide, apex acuminate, base angustate or sub-rotundate, 3–5-nerved, glabrous on both surfaces, margins papillose, with a short petiole *c.* 5mm long. Flowers 1–5 in a raceme or sub-umbelliformly disposed, nodding; bracts leaf-like, ovate, 3.5–4cm long, 2–2.5cm wide; pedicels up to 11cm long; perianth segments 6–7.5cm long, recurved, margins undulate, white, with purple-red blotches and spots on the lower third to half, nectaries with red fimbriate projections and papillae on each side; stamens spreading outwards, filaments 5.5–6cm long, green, glabrous, anthers 1.5–1.8cm long, crimson; ovary cylindric, *c.* 1.5cm long; style twice as long as the ovary, stigma swollen, shallowly trilobed. Capsule sub-globose, 3cm broad, pale brown, pedicel swollen at maturity.

Fig. 10.
Lilium speciosum Thunb. var. *gloriosoides* Baker.

Flowering period: July–August. Fruiting period: October.

Occurs in Anhui, Jiangxi, Zhejiang, Hunan, Guangxi and Taiwan. It grows in damp, shady woodland and among herbs on mountain slopes, at 650–1000 metres above sea level. The bulb is used as medicine, and can be eaten. The flowers are very beautiful, and this is a well-known ornamental plant.

† The Chinese variety of *Lilium speciosum* has often been considered to be the most beautiful form of the species, with brighter red spots and very elegant flower-form. It is unfortunately highly unlikely that any authentic stocks of it remain in cultivation in the west. In Taiwan it is apparently not very common, and is restricted to the northern foothills of the Taiwan mountains, in the vicinity of Pinglin (Heirinbi), about 20 miles south-east of Taibei. It is recorded as growing there on wet sandstone cliffs or among herbs and low shrubs.

Its type locality is, however, the Lu Shan mountains in Jiangxi, on the mainland. It was first collected there by Père David in 1868. E. H. Wilson 'searched for it in vain' when he visited the range in 1907, but I am pleased to say that I can confirm that it still grows there. I even found it to be locally quite abundant when I stayed for several days in Lu Shan in late July 1982. It was not, however, in flower at that time, though the buds would probably have started to open within the next couple of weeks. It grew at the edges of woodland, in quite heavy shade, on the middle to upper slopes of the mountains between about 900 and 1200 metres above sea level. To judge from associated vegetation, it occurred only where the soil was acid (there is considerable variation in soil-type in the Lu Shan mountains).

It is not, however, as has often been stated by previous authors, confined to these two locations. I have seen specimens from Jiu Hua Shan in Anhui and Tian Tai Shan in Zhejiang, and Liang Sung-yun also gives Hunan and Guangxi among the provinces in which it occurs. It would therefore seem to be rather widely distributed, though probably it is only found in a few isolated mountain sites in each province.

VI. Section Dimorphophyllum S. G. Haw, sect. nov. in Appendix 1.

19 (23). **Lilium henryi** Baker in *Gard. Chron.* ser. 3, 4: 660. 1888; et in *Bot. Mag.* t. 7177. 1871; Hu et Chun, *Ic. Pl. Sin.* 1: 48, t. 48. 1927; Woodc. et Stearn, *Lil. World* 226, f. 60. 1950.

Bulb sub-globose, 5cm tall, 7cm in diameter; scales oblong, acute at the apex, 3.5–4.5cm long, 1.4–1.6cm wide, white. Stem 100–200cm tall, streaked with purple, glabrous. Leaves dimorphous, the middle and lower ones oblong-lanceolate, 7.5–15cm long, 2–2.7cm wide, apex acuminate, base sub-rotundate, 3–5-nerved, glabrous on both surfaces, entire, petioles *c.* 5mm long; upper ones ovate, 2–4cm long, 1.5–2.5cm wide, apex acute, base sub-rotundate, sessile. Raceme with 2–12 flowers; bracts ovate, leaf-like, 2.5–3.5cm long, apex acute; pedicels 5–9cm long, spreading horizontally, each pedicel usually bearing 2 flowers; perianth segments lanceolate, recurved, orange, sparsely spotted with black, 5–7cm long, up to 2cm wide, entire, nectaries with numerous fimbriate projections on each side; stamens spreading outwards, filaments subulate, 4–4.5cm long, glabrous, anthers deep vermilion; ovary sub-cylindric, 1.5cm long; style 5cm long, stigma somewhat swollen, shallowly trilobed. Capsule oblong, 4–4.5cm long, *c.* 3.5cm broad, brown.

Occurs in Hubei, Jiangxi and Guizhou. It grows on mountain slopes at 700–1000 metres above sea level.

This species is very like *L. rosthornii* Diels, both have dimorphous leaves; but they differ in that this species has oblong-lanceolate leaves, ovate bracts and an oblong capsule 4–4.5cm long.

† *Lilium henryi* was introduced to western gardens through collections made by Augustine Henry in the limestone gorges near Yichang, on the Yangtze River in Hubei province. The first bulbs flowered at Kew in 1889. E. H. Wilson also sent large shipments of bulbs to both Britain and the USA. This

1
2
3
4
5
6
7
8
9

has proved to be one of the easiest of Chinese lilies to grow, and it remains common in cultivation, despite the popularity of hybrids. It is particularly useful for gardeners with chalky soils, for it is one of the few Chinese lilies with a definite dislike for acid conditions. In cultivation its bulbs may grow to as much as 20cm in diameter, and they should be planted with a good depth of soil over them to allow the stem to make plenty of roots. It often grows very tall, and is rather weak-stemmed, so that it is a good idea to plant it where it may grow through shrubs and gain support from their branches.

It has been hybridized with several species, including lilies of section Regalia and *Lilium speciosum*, and usually seems to have endowed its offspring with its own strength of constitution.

20 (24). **Lilium rosthornii** Diels in *Bot. Jahrb.* 29: 243. 1901.

Bulb not seen. Stem 40–100cm tall, glabrous. Leaves scattered, those on the middle and lower parts of the stem linear-lanceolate, 8–15cm long, 8–10mm wide, apex acuminate, narrowed at the base into a short petiole, glabrous on both surfaces, entire; those on the upper part of the stem ovate, 3–4.5cm long, 10–12mm wide, apex acute, base angustate, midrib prominent, glabrous on both surfaces, entire. Raceme with up to 9 flowers, or rarely bearing a solitary flower; bracts broadly ovate, 3–3.5cm long by 1.5–2cm wide, apex acute, base angustate; pedicels (3–) 7–8cm long; perianth segments recurved, yellow or orange, with purple-red spots, 6–6.5cm long, 9–11mm wide, entire, nectaries with many fimbriate projections on each side; stamens spreading outwards, filaments *c.* 6–6.5cm long, glabrous, anthers 1.2–1.4cm long; ovary cylindric, 1.5–2cm long, *c.* 2mm broad; style 4–4.5cm long, stigma somewhat swollen. Capsule long-oblong, 5.5–6.5cm long, 1.4–1.8cm broad, olive-green.

Occurs in Sichuan, Hubei and Guizhou. It grows in mountain ravines, by streams or in woodland, at 350–900 metres above sea level. The bulb can be used as medicine, or eaten.

This species is very similar to *L. henryi* Baker, both have dimorphous leaves; but they differ in that the middle and lower leaves of this species are longer and linear-lanceolate, and the capsule is long-oblong, 5.5–6.5cm long.

† This lily is known in the west from only one or two collections. Diels's type specimen came from near Nanchuan in Sichuan, south-east of Chongqing and not far from the border with Guizhou province. I have also seen a specimen

Fig. 11.
1–5 *Lilium henryi* Baker: 1 Upper part of plant; 2 Middle part of plant; 3 Fruit; 4 outer perianth segment; 5 Inner perianth segment.
6–9 *L. rosthornii* Diels: 6 Leaf from middle part of plant; 7 Leaves from upper part of plant; 8 Bract; 9 Fruit.

1
2
3
4
5
6
7
8

of what I consider to be this species which was collected near Badong in western Hubei, in the area of the Yangtze gorges. The flowers are more heavily spotted than those of *L. henryi*, and the spots are much redder in colour. This would seem to be quite a distinct and desirable lily. It unfortunately appears to be a rare plant in the wild.

VII. Section Sinomartagon Comber

21 (18). **Lilium nepalense** D. Don in *Mem. Wern. Soc.* 3: 412. 1821.

Bulb sub-globose, *c.* 2.5cm tall, *c.* 2cm in diameter; scales lanceolate or ovate-lanceolate, 2–2.5cm long, 1–1.2cm wide, white. Stem 40–120cm tall, papillose. Leaves scattered, lanceolate or oblong-lanceolate, 5–10cm long, 2–3cm wide, apex acuminate, base angustate, margins papillose, 5-nerved, glabrous on both surfaces. Flowers solitary or 3–5 in a raceme; bracts oblong-lanceolate, 5.5–10cm long; pedicels 9–13cm long; perianth pale yellow or greenish-yellow, tinged with purple in the throat, somewhat trumpet-shaped, perianth segments revolute, 6–9cm long, 1.6–1.8cm wide, nectaries not papillose; filaments 5–5.5cm long, glabrous, anthers 8–9mm long; ovary cylindric, 1.5–1.8cm long, style 4–5cm long, stigma swollen, *c.* 4mm in diameter.

Flowering period: June–July.

Occurs in southern Tibet (Gyirong and Nyalam) and Yunnan (western part). It grows among shrubs in mixed woodland and by roadsides, at 2650–2900 metres above sea level. It is also distributed in Nepal, Bhutan and India.

var. **burmanicum** W. W. Sm. in *Trans. Bot. Soc. Edinb.* 28: 135. 1922.

This variety differs from the type in having long, narrow leaves, 9–16cm long, 0.8–1.4cm wide.

Occurs in Yunnan. It grows on grassy slopes or at the edges of woods, at 1200–2200 metres above sea level. It is also distributed in Burma and Thailand.

Fig. 12.
1–4 *Lilium nepalense* D. Don var. *burmanicum* W. W. Sm.: 1 Upper part of plant; 2 Outer perianth segment; 3 Inner perianth segment; 4 Stamen and pistil.
5–8 *L. nepalense* D. Don var. *ochraceum* (Franch.) Liang: 5 Leaf; 6 Outer perianth segment; 7 inner perianth segment; 8 Stamen and pistil.

var. **ochraceum** (Franch.) Liang, *Flora RPS* 14: 138. 1980. – *Lilium ochraceum* Franch. in *Journ. de Bot.* 6: 319. 1892.

This variety differs from the type in having shorter and narrower leaves, 3–5.5cm long, 8–10mm wide; and in having ovate bracts with acute tips, 3.5cm long, 1.1cm wide.

Flowering period: July–August. Fruiting period: October–November.

Occurs in Yunnan (north-west and north-east parts), Sichuan (Xichang) and Guizhou (Anshun). It grows on grassy slopes or among shrubs, at 2000–3000 metres above sea level.

† This is a very variable species which presents many problems to the taxonomist. These have been reviewed in Chapter 4, where I give my reasons for concurring with the above treatment of *Lilium nepalense* and its varieties. The type variety is now the best known in the west, and is the only one still in cultivation, having been collected on numerous occasions in recent years from the Nepal Himalaya. It ranges all along the Himalayan range from a western limit in Uttar Pradesh to as far east as western Yunnan (according to the *Flora RPS*), occurring within the borders of Tibet only on the wet southern slopes of Mt. Xixabangma, in the vicinity of Gyirong and Nyalam. Var. *burmanicum* occurs in southern Yunnan, Burma and Thailand, and var. *ochraceum* ranges across north-west Yunnan, through the Xichang area of Sichuan and north-east Yunnan to western Guizhou (it appears to be rare in the eastern half of its range). There does, however, seem to be at least some overlap in the range of the three varieties in western Yunnan, where intermediate forms also occur. More studies are needed of the forms of *L. nepalense* from this area to settle finally the taxonomic difficulties associated with this species.

None of the varieties of this lily are easy to cultivate. All have, at various times and from several collections, been grown in western gardens, but now only the type variety remains, and only because it has been frequently re-introduced from Nepal. Undoubtedly the main difficulty in growing *L. nepalense* lies in keeping it sufficiently dry in winter, and many growers have found it easier to keep alive under glass. It is currently cultivated in the Temperate House at Kew, for example, from seed collected in Nepal in 1977. The minimum winter temperature is 7° C., and in these frost-free conditions absolute dryness does not seem to be necessary. Other growers recommend planting in tubs outdoors, which are kept moderately shaded and covered from early October to April to keep out rain. But it is certainly possible to succeed with this lily in the open garden, both in the south of England and in Scotland. I have recently seen *L. nepalense* growing well and flowering in the peat garden at RBG Edinburgh. It prefers an acid soil, and must have very free drainage, though with plenty of moisture in summer. Most growers also recommend applying plenty of fertilizer in spring, or giving regular liquid feeds while the plants are in leaf. As with all plants, there can be no single

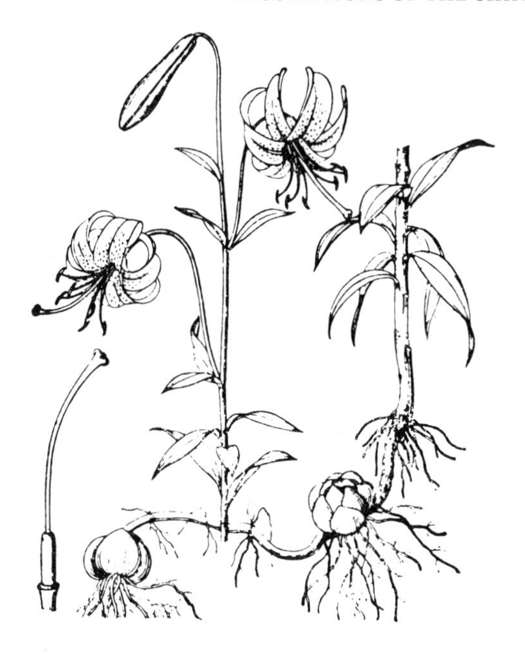

Fig. 13.
Lilium wardii Stapf ex Stearn.

recipe for success, but recent experience with this lily should encourage growers to give it a try.

22 (19). **Lilium wardii** Stapf ex Stearn in *Journ. Roy. Hort. Soc.* 57: 291. 1932; Woodc. et Stearn, *Lil. World* 362, f. 118. 1950.

Bulb sub-globose, 2–3cm tall, 2.5–4cm in diameter; scales ovate, 1.5–2cm long, 7–9mm wide. Stem 60–100cm tall, purplish-brown, papillose. Leaves scattered, narrowly lanceolate, 3–5.5cm long, 6–7mm wide, with three obvious veins impressed on the upper surface, glabrous on both surfaces, margins papillose. Flowers 2–10 in a raceme, or rarely solitary, nodding; bracts leaf-like, ovate to lanceolate, 2.5–4.5cm long, 5–16mm wide; perianth segments revolute, pale purple-red or pink, spotted with deep purple, oblong or lanceolate, 5.5–6cm long, 8–10mm wide, nectaries without fimbriate projections on each side; filaments subulate, 4–4.5cm long, glabrous, anthers purple, pollen orange; ovary cylindric, *c.* 1cm long; style three times or more longer than the ovary, stigma sub-globose, trilobed.

Flowering period: July.

Occurs in Tibet. It grows in grassy places or among shrubs on mountain slopes, at 2000–3000 metres above sea level.

† This beautiful lily is named after its discoverer, Frank Kingdon Ward, who first collected it in 1924. It was initially confused with *Lilium duchartrei*, and

not separately described until 1932. In 1930 it was awarded an FCC, and has remained in cultivation ever since despite high susceptibility to virus disease. Its usually stoloniform stems, which produce offset bulblets, and the abundance with which it sets seed, allow it to be readily propagated. Seed should always be used for preference, to reduce the risk of virus infection. It seems to be mainly because of disease, and because its bulbs are not individually very long-lived, that *Lilium wardii* has failed to become more common in cultivation. It is not one of the most difficult lilies to grow, being moderately tolerant of soil conditions, providing that there is high humus content and good drainage. Some growers recommend mulching with pine needles. It has been found growing among pine trees in the wild, though it is not confined to such situations.

Its natural range is not very large. *L. wardii* grows only in south-east Tibet, in the valleys of the Yarlung Zangbo and Zayü Rivers and their tributaries. Its western limit is in the vicinity of Nyingchi, at about 94° E., and eastward it extends to beyond Zayü, almost to the border with Yunnan.

23 (20). **Lilium stewartianum** Balf. f. et W. W. Sm. in *Trans. Bot. Soc. Edinb.* 28: 127, t. 4. 1922; Woodc. et Stearn, *Lil. World* 343, f. 106. 1950.

Bulb ovoid, 2cm tall, 2cm in diameter; scales ovate-lanceolate, white. Stem 20–50cm tall, green, sometimes spotted with purple-red, glabrous. Leaves scattered, linear, 2.5–7cm long, 3–4mm wide, midrib slightly prominent, margins sparsely papillose. Flower solitary, fragrant, greenish-yellow, spotted with deep red, nodding; perianth segments oblanceolate-oblong, 4.5–5cm long, 7–9mm wide, revolute towards the tip, nectaries without fimbriate projections on each side; filaments subulate, 3cm long, glabrous; ovary cylindric, 2–2.2cm long, *c.* 3mm broad, purple; style as long as the ovary, stigma capitate. Capsule oblong or ellipsoid, 2–2.5cm long, 1.5–2cm broad, brown.

Flowering period: July–August. Fruiting period: October.

Occurs in Yunnan (north-west part). It grows on limestone rocks or open, rocky grassland, or at the edges of woods, at 3600–4300 metres above sea level.

This species is similar to *L. fargesii* Franch., but differs in that the latter bears its flowers in a raceme and has cristate projections on each side of the nectaries.

† *Lilium stewartianum* has been in cultivation in both Britain and the USA, from collections made by Joseph Rock, but is only recorded to have flowered once, in 1952. It has not been grown since. It seems to be rather uncommon in the wild, and has only a limited range of distribution, from the mountains of the Lijiang area northwards to the Zhongdian plateau. It was first collected by George Forrest in 1913.

It is close not only to *Lilium fargesii*, but also to *L. nepalense* and *L. taliense*.

Its perianth segments form a long tube before reflexing, and in colour resemble those of *L. nepalense*, though it is clearly distinguished by its smooth stem and very narrow leaves. It is a very high altitude lily, growing at higher levels than any other species of this section.

24 (21). **Lilium taliense** Franch. in *Journ. de Bot.* 6: 319. 1892; Woodc. et Stearn, *Lil. World* 348, f. 111. 1950. – *L. feddei* Lévl. in *Rep. Sp. Nov. Fedde* 11 : 303. 1912.

Bulb ovoid, *c.* 3cm tall, 2.5cm in diameter; scales lanceolate, 2–2.5cm long, 5–8mm wide, white. Stem 70–150cm tall, sometimes purple-spotted, papillose. Leaves scattered, linear or linear-lanceolate, 8–10cm long, 6–8mm wide, midrib prominent, glabrous on both sides, margins papillose. Flowers 2–5, or rarely up to 13, in a raceme, nodding; bracts leaf-like, 3–5cm long, 4–8mm wide, margins papillose; perianth segments revolute, oblong or oblong-lanceolate, 4.5–5cm long, *c.* 1cm wide, white with purple spots, inner segments slightly wider than outer segments, nectaries without fimbriate projections on each side; filaments subulate, *c.* 3cm long, glabrous; ovary cylindric, 1.4–1.6cm long, 3–4mm broad; style as long as or slightly longer than the ovary, stigma capitate, trilobed. Capsule oblong, 3.5cm long, 2cm broad, brown.

Flowering period: July–August. Fruiting period: September.

Occurs in Yunnan and Sichuan. It grows on grassy mountain slopes or in woodland, at 2600–3600 metres above sea level.

† This is a striking species, taller than the rather similar *L. duchartrei*, which also has white flowers. It was known from pressed specimens as early as the 1880s, but only came into western gardens in 1935. Cultivated stocks have never been very large, though it has grown well at the Oregon Bulb Farms and been widely distributed from there. It will grow on moderately alkaline soils, and is probably not excessively difficult to manage, but is not self-fertile and will therefore not produce seed unless two distinct clones are cross-pollinated. In Britain at least, seed seems rarely to be available, perhaps because most growers have stock which has been vegetatively propagated from one original source.

Its natural range of distribution is from the mountains between the Lancang Jiang (Mekong) and Jianchuan in north-western Yunnan, north-eastwards through the Lijiang and Yongning areas to the district of Muli in Sichuan, and as far east as Yao Shan (Io chan) near Qiaojia in north-east Yunnan.

25 (25). **Lilium duchartrei** Franch. in *Nouv. Arch. Mus. Paris* ser. 2, 10: 90 (Pl. David 2: 128) 1887; et in *Journ. de Bot.* 6: 316. 1892; Woodc. et Stearn, *Lil. World* 212, f. 43. 1950. – *L. lankongense* Franch. in *Journ. de Bot.* 6: 317. 1892. – *L. forrestii* W. W. Sm. in *Notes Bot. Gard. Edinb.* 8: 192. 1914. – *L. farreri* Turrill in *Gard. Chron.* ser 3, 66: 76. 1919; et in *Bot. Mag.* t. 8847. 1920.

Bulb ovoid, 1.5–3cm tall, 1.5–4cm broad, stoloniferous; scales ovate to broadly lanceolate, 1–2cm long, 0.5–1.8cm wide, white. Stem 50–85cm tall, streaked with pale purple. Leaves scattered, lanceolate to oblong-lanceolate, 4.5–5cm long, *c.* 1cm wide, glabrous on both sides, 3–5-nerved, sometimes with papillose margins. Flowers solitary or several in a raceme, or in corymbose or umbelliform-racemose arrangement, nodding, fragrant; bracts leaf-like, lanceolate, 2.5–4cm long, 4–6mm wide; pedicels 10–22cm long; perianth white or pink, spotted with purple; perianth segments revolute, 4.5–6cm long, 1.2–1.4cm wide, nectaries with papillae on both sides; filaments 3.5cm long, glabrous, anthers narrowly oblong, *c.* 1cm long, yellow; ovary cylindric, 1.2cm long, 1.5–3mm broad; style at least twice as long as ovary, stigma swollen. Capsule ellipsoid, 2.5–3cm long, *c.* 2.2cm broad. Seeds flattened, with wings 1–2mm wide.

Flowering period: July. Fruiting period: September.

Occurs in Sichuan, Yunnan, Tibet and Gansu. It grows in alpine meadows, at the edges of woods, or in thickets, at 2300–3500 metres above sea level.

According to the accounts of Wilson, Woodcock and Stearn, and others, *L. lankongense* Franch. differs from this species, having flowers arranged in a raceme, leaves very crowded, nerves prominent on the undersurfaces of the leaves, pink perianth segments, comparatively narrow seed-wings (*c.* 1mm wide) and other characteristics. But examination of the specimens kept at the Botanical Institute of Academia Sinica shows that the inflorescence varies greatly, from a solitary flower to a raceme or almost an umbel, sometimes apparently being umbelliform, but with pedicels of unequal lengths, sometimes with a leaf-like bract on the middle of the pedicel, while the inflorescence is still really a raceme; as for the other features, such as whether the leaves are crowded or not, the prominence of the nerves on the undersides of the leaves and the colour of the flowers, these are all variable, with intermediate and overlapping forms. The only difference is the breadth of the seed-wings, but even here there is overlap. Because of this, we include *L. lankongense* Franch. in this species.

There is also another species similar to this one, *L. ninae* Vrishcz. (1968), which occurs in Tibet. It differs in having linear leaves, usually one pair of bracts, and no stolons. We have as yet not seen any specimens, so it is difficult to determine its status.

† It will seem strange to those familiar with *L. duchartrei* and *L. lankongense* in gardens that they should be reduced to synonymy. But as already explained in Chapter 4, our garden stocks derive from collections made near the northern and southern limits of the range of distribution of this lily, and represent extremes of variation within what I am convinced must be regarded as one species. In the large intervening area of distribution, all kinds of intermediate forms occur. Nor is there any geographic separation between *L. duchartrei* and *L. lankongense*, as some authors have asserted. I have seen one herbarium sheet, Kingdon Ward 4339 (at Edinburgh), on

which there are two plants collected near Muli in Sichuan, one of which has a racemose and the other an umbellate inflorescence. There are also specimens of both forms of this lily from within a fairly small area (between 27° 34′ and 28° 12′ N.) on the mountains of the Lancang Jiang-Nu jiang (Mekong-Salween) divide in north-west Yunnan, and from the Lijiang area.

The total range of this species extends from a southern limit of about 26° 30′ N. in north-west Yunnan, across western Sichuan to southern Gansu, as far as about 35° N. In the south-west of this range, it extends into the Zayü region of Tibet, and in the east reaches Wa Shan and the Songpan area in Sichuan, and south-east Gansu.

Both of the cultivated forms have persisted well since being introduced in 1915 (from Gansu) and 1920 (from Yunnan). The Gansu form (*L. duchartrei* of gardens) was regarded as rather an easy lily until quite recently, but it has now become much less common in British gardens. This may well be because it has been commonly propagated by means of the numerous offset bulblets produced on its stoloniferous underground stems, rather than by seed, and has therefore finally succumbed to a build-up of disease in cultivated stocks. Growers should take any opportunity which arises to grow uninfected plants from seed, for it would be a great pity to lose this lily. The Yunnan form (*L. lankongense* of gardens) was conversely always considered moderately difficult (except perhaps in Scottish gardens), but now seems to be more commonly grown than the form from Gansu. There is no question but that both forms prefer the climate of the northern half of Britain to that of the south. Both are moderately lime-tolerant, and prefer a humus-rich, well-drained, loamy soil. They need plenty of water during the growing season, but must be reasonably dry in winter. They will stand full exposure to sunlight, but prefer to have their roots kept cool. In the south of England they are best given some shade and protection by being planted among dwarf shrubs or with ground-cover of suitable alpine plants.

26 (29). **Lilium davidii** Duchartre in Elwes, *Monogr. Lil.* t. 24. 1877; Wilson, *Lil. East. As.* 81, t. 13. 1925. – *L. cavaleriei* Lévl. et Vnt. in *Lilac. etc. Chine* 44. 1905. – *L. thayerae* Wilson in *Kew Bull.*, 266. 1913.

Bulb flattened-globose or broadly ovoid, 2–4cm tall, 2–4.5cm in diameter; scales broadly ovate to ovate-lanceolate. 2–3.5cm long, 1–1.5cm wide, white. Stem 50–100cm tall, sometimes tinged with purple, densely papillose. Leaves numerous, scattered, comparatively densely crowded in the middle part of the stem, linear, 7–12cm long, 2–3 (–6)mm wide, apex acute, margins revolute and also conspicuously papillose, midrib obvious, frequently impressed on the upper surface and prominent on the lower surface, leaf-axils with white woolly hairs. Flowers solitary or 2–8 arranged in a raceme, nodding; bracts leaf-like, 4–7.5cm long, 3–7mm wide; pedicels 4–8cm long; perianth orange, with dark purple spots on the basal two-thirds approximately; outer segments 5–6cm long, 1.2–1.4cm wide; inner segments slightly wider than the outer ones, nectaries papillose, and with a few fimbriate projections at their outer edge; filaments 4–5.5cm long, glabrous, anthers 1.4–1.6cm long, pollen deep tangerine-colour; ovary cylindric, 1–1.2cm long, 2–3mm broad; style at least twice as long as the ovary, stigma swollen, shallowly trilobed. Capsule long-oblong, 3.5cm long, *c.* 1.6–2cm broad.

Flowering period: July–August. Fruiting period: September.

Occurs in Sichuan, Yunnan, Shaanxi, Gansu, Henan, Shanxi and Hubei. It grows on mountain slopes in meadows, moist places in woodland, or at the edges of woods, at 850–3200 metres above sea level. The bulb contains starch and is of good quality, gives high yields in cultivation, and may be used as food.

† *Lilium davidii* is well-known in cultivation, and is one of the easier Chinese lilies. A number of varieties are usually recognized, such as var. *willmottiae* (Wilson) Raffill, but these are not mentioned in the above account, probably because they intergrade and are not easily distinguished in the wild. This is a very widespread lily, common throughout the mountainous regions of western China from central Yunnan across eastern and western Sichuan to western Hubei and southern Gansu and Shaanxi. It also extends along the Qinling range into Henan province, and northwards at least as far as Huo Shan (Lao Ye Ding) in southern Shanxi. In the Tengchong, Dali and Lijiang regions of Yunnan it has long been cultivated for its edible bulbs, and this practice is now spreading, mainly to fulfil demand for bulbs for medicinal use.

This lily has been much used in the west as a parent of hybrids, and has given rise to several important strains, including the Preston Hybrids from Canada and the Fiesta Hybrids from the Oregon Bulb Farms. It has, in fact, been to a considerable extent replaced in gardens by its hybrid offspring, and is now much less commonly grown. But it is easily propagated from seed (as well as producing offset bulblets), is tolerant of a range of soils and of drought, and is reasonably resistant to disease. It is also one of the more elegant orange Sinomartagons. There are therefore many reasons for it to be restored to popularity.

27 (26). **Lilium leichtlinii** Hook. f. in *Bot. Mag.* t. 5673. 1867.

Occurs in Japan, but does not occur in China.

var. **maximowiczii** (Regel) Baker in *Gard. Chron.* 1422. 1871; Wilson, *Lil. East. As.* 71, t. 11. 1925. – *L. maximowiczii* Regel in *Gartenfl.* 17: 322, t. 596. 1868.

Bulb globose, 4cm tall, 4cm broad, white. Stem 0.5–2m tall, spotted with purple, papillose. Leaves scattered, narrowly lanceolate, 3–10cm long, 0.6–1.2cm wide, margins papillose, upper leaf axils not bulbiliferous. Flowers 2 or 3–8 arranged in a raceme, rarely solitary, nodding; bracts leaf-like, lanceolate, 5–7.5cm long, 8mm wide; pedicels rather long, (3.5–) 10–13cm long; perianth segments revolute, red, spotted with purple, 4.5–6.5cm long, 0.9–1.5cm wide, nectaries papillose and also with fimbriate projections on each side; stamens divergent, filaments 3.5–4cm long, glabrous, anthers 1.1cm long, vermilion; ovary cylindric, 1.2–1.3cm long, 2–3mm broad, style 3cm long.

Flowering period: July–August.

Occurs in Shaanxi, North China and Manchuria. It grows in sandy places in valley bottoms, at up to 1290 metres above sea level. It is also distributed in Korea, Japan and the USSR.

L. leichtlinii var. *maximowiczii* is very like *L. lancifolium* Thunb., but differs in not having bulbils in the upper leaf axils, and in having red flowers with purple spots.

† This is generally stated to be an easy lily to grow, yet it is not often seen in cultivation and is not very readily obtainable. It has undoubtedly suffered because it is so similar to *L. lancifolium* but less robust and without the bulbils that make propagation of the Tiger Lily such an easy matter. It does seem to be more variable in flower colour than the Tiger Lily, from orange to various shades of red, and will set viable seed, which cultivated *L. lancifolium* rarely does. It also commonly flowers rather late, often in September. It therefore ought to be well worth growing, and was indeed awarded an FCC in 1872.

The distribution within China given for this lily in the *Flora RPS* is much wider than has been previously recorded. In fact, many recent western writers have not included China in its distribution range at all. There is no doubt, however, that it occurs wild in Manchuria, and I have seen specimens collected near the Yalu River and in the Changbai Shan range. It has been long cultivated in China, and it may be that its range has been extended by the agency of man. It is also likely, however, that it was often overlooked by western collectors because it is so similar to *L. lancifolium*.

E. H. Wilson saw *Lilium leichtlinii* var. *maximowiczii* growing wild in central Korea, 'in black mould alongside streams and ponds, in swampy places, and on slopes, always among coarse herbs and bushes', and in Japan on the lower slopes of an active volcano. It would appear to enjoy plenty of moisture in the soil, but not to be confined to very wet situations. Volcanic soils are usually poor in nutrients and very free draining.

Var. *maximowiczii* is the wild form of the yellow-flowered *L. leichtlinii*, known from cultivation in Japan and probably occurring naturally there. The yellow form is no more than an abnormal colour variation, however, and it is only because it became known to western botanists before its common wild variety that it came to be the type. It was accidentally introduced to Britain with a consignment of *L. auratum*, flowered with Messrs. Veitch in 1867, and was described in the same year. Regel described his *Lilium maximowiczii* in a work dated 1866, so that some authorities have taken this to be the prior name; it appears, however, that his description was not actually published until 1868.

28 (34). **Lilium lancifolium** Thunb. in *Trans. Linn. Soc.* 2: 333. 1794. – *L. tigrinum* Ker-Gawl. in *Bot. Mag.* t. 1237. 1810.

Bulb broadly sub-globose, *c.* 3.5cm tall, 4–8cm in diameter; scales broadly ovate, 2.5–3cm long, 1.4–2.5cm wide, white. Stem 0.8–1.5m tall, streaked with purple, white-lanate. Leaves scattered, oblong-lanceolate or lanceolate, 6.5–9cm long, sub-glabrous on both surfaces, with white woolly hairs at the apex and papillose margins, 5–7-nerved, upper leaf axils bulbiliferous. Flowers 3–6 or more, nodding; bracts leaf-like, ovate-lanceolate, 1.5–2cm long, 2–5mm wide, apex blunt and white-lanate; perianth segments lanceolate, revolute, vermilion, spotted with dark purple; outer segments 6–10cm long, 1–2cm wide; inner segments slightly wider, nectaries papillose and with fimbriate projections on each side; stamens divergent, filaments 5–7cm long, light red, glabrous, anthers oblong, *c.* 2cm long, 2–3mm broad; style 4.5–6.5cm long, stigma somewhat swollen, trilobed. Capsule narrowly long-ovate, 3–4cm long.

Flowering period: July–August. Fruiting period: September–October.

Occurs in Jiangsu, Zhejiang, Anhui, Jiangxi, Hunan, Hubei, Guangxi, Sichuan, Qinghai, Tibet, Gansu, Shaanxi, Shanxi, Henan, Hebei, Shandong and Jilin. It grows on mountain slopes among shrubs, in meadows, by the sides of roads or on the banks of rivers, at 400–2500 metres above sea level, and is widely cultivated. It is also distributed in Japan and Korea. The bulb has a high starch content and can be eaten or used medicinally; the flowers contain essential oil and can be used in making perfume.

† The orange Tiger Lily must be the most familiar Chinese lily to western gardeners. It has been in cultivation in Britain since 1804, and seems to have remained quite popular despite the competition from hybrids. Virtually all the cultivated stocks are triploid, and therefore rarely set fertile seed, but the bulbils which are plentifully produced in the leaf axils of this lily make propagation easy. Although it is often infected with virus diseases, it rarely shows symptoms. Care should be taken, however, where susceptible species are also being grown, to keep this lily apart and to watch carefully for the aphids which spread the infection.

This is an easy lily to grow, but will not tolerate too much lime in the soil and is averse to heavy shade. Plenty of humus and an occasional feed will encourage it to grow tall and produce plenty of flowers. The bulbils may be removed as soon as they reach a reasonable size, planted in pots and grown on for a couple of seasons. They may then be transferred to the open garden, and should reach flowering size in about three years.

It will be regretted by most gardeners that the formerly accepted name of *L. tigrinum* is rejected by the *Flora RPS*, but since the *Flora Europea* decided to use Thunberg's name *L. lancifolium* for this lily (which is naturalized in Austria), an increasing number of botanical authorities have followed this lead. There is no doubt that Thunberg's name is earlier, but his description is so poor that for many years it was considered uncertain which lily he intended. There are no good arguments, however, for continuing to use two names for the same plant, and as Thunberg's name has now been accepted by at least two major *Floras* it seems pointless to try to maintain any other.

The Tiger Lily has not been cultivated in China for nearly so long as has commonly been stated by previous western authors (see Chapter 3). It therefore seems to me to be most unlikely that its distribution in China has been greatly affected by the agency of man. In fact, it is at least as probable that widespread digging of its bulbs from the wild for use as food or medicine has made the Tiger Lily less common. But it is certainly still wild on the Lu Shan range in Jiangxi, where I have seen flowering stems picked from the mountainsides by local Chinese, and several provincial or regional Chinese *Floras* record it as wild as well as cultivated. There can be no doubt that it is naturally a widely occurring lily in China.

Studies of wild populations in Japan have shown that both diploid and triploid forms of this lily occur together in the same localities. The triploid form (2n = 36) is more common, but the diploid form (2n = 24) has been found at no less than seven locations on Tsushima Island, and the two forms were morphologically indistinguishable. There is no reason to suppose that the triploid forms of the Tiger Lily are of hybrid origin, as has often been suggested. It would be useful to obtain wild diploid forms of *L. lancifolium*, as they should produce seed much more readily than the present triploid cultivated stocks.

29 (27). **Lilium pumilum** Delile in Redouté, *Liliac.* 7: t. 378. 1812; Woodc. et Stearn, *Lil. World* 324, f. 97. 1950. – *L. tenuifolium* Fisch. in *Cat. Jard. Gorenki* 8. 1812, nomen nudum; Hook. in *Bot. Mag.* t. 3410. 1832. – *L. potaninii* Vrishcz. in *Bot. Journ. URSS* 53: 1472. 1968.

Bulb ovoid or conoidal, 2.5–4.5cm tall, 2–3cm in diameter; scales oblong or long-ovate, 2–3cm long, 1–1.5cm wide, white. Stem 15–60cm tall, papillose, sometimes streaked with purple. Leaves scattered on the middle section of the stem, linear, 3.5–9cm long, 1.5–3cm wide, the midrib prominent on the undersurface, margins papillose. Flowers solitary or several in a raceme, bright red, usually unspotted, occasionally with a few spots, nodding; perianth segments revolute, 4–4.5cm long, 0.8–1.1cm wide, nectaries papillose; filaments 1.2–2.5cm long, glabrous, anthers long-ellipsoid, *c.* 1cm long, yellow, pollen almost red; ovary cylindric, 0.8–1cm long; style a little longer than to more than twice as long as the ovary, 1.2–1.6cm long, stigma swollen, 5mm in diameter, trilobed. Capsule oblong, 2cm long, 1.2–1.8cm broad.

Flowering period: July–August. Fruiting period: September–October.

1
2
3
4
5
6
7
8
9

Occurs in Hebei, Henan, Shanxi, Shaanxi, Ningxia, Shandong, Qinghai, Gansu, Inner Mongolia, Heilongjiang, Liaoning and Jilin. It grows on grassy mountain slopes or at the edges of woods, at 400–2600 metres above sea level. It is also distributed in the Soviet Union, Korea and Mongolia. The bulb contains starch and is edible, and can also be used as medicine, effective as a tonic, an antitussive expectorant, diuretic etc. The flowers are beautiful, and this plant may be cultivated for ornament; they also contain a volatile oil, which can be extracted for use as perfume.

Before the perianth segments become revolute this species is difficult to distinguish from *L. concolor* Salisb., but it has large flowers, with perianth segments 4–4.5cm long, and the style is a little longer than, to more than twice as long as, the ovary; while *L. concolor* has smaller flowers, with perianth segments 2.2–3.5cm long, and the style is shorter than the ovary.

† This is the most common lily of north China, occurring in every province and autonomous region through which the Yellow River flows and northward to beyond the borders. In much of this range it is fairly frequent, even in the vicinity of major cities. I have seen quite a number of plants, for example, in the hills near the western suburbs of Beijing, both in the vicinity of Wo Fo Si (the Sleeping Buddha Temple), and of Xiang Shan (Fragrant Hills) Park. It is a plant of hot, dry, rocky slopes, usually growing among moderately dense vegetation of herbs and shrubs, sometimes among trees, but rarely very heavily shaded. In the hills west of Datong in northern Shanxi I have seen it growing on very dry, sparsely vegetated hillsides. It varies considerably in height and number of flowers according to the conditions in which it grows. All the plants seen near Datong were small (less than 30cm) and with just a solitary flower, while those near Beijing were much taller (up to almost a metre) and bore four or five flowers per stem.

In cultivation *L. pumilum* is quite easily grown in a dry, sunny situation, but is rarely very long-lived and should be regularly propagated if it is to be maintained. It usually sets plenty of seed and can be grown to flowering size in about three years. Its bright scarlet flowers are very showy and make a strong splash of colour when they open in early summer. There are also yellow and white forms of this lily, which are much rarer both in the wild and in cultivation. The yellow forms are very attractive and desirable, and at least some of them come true from seed.

Recently it has been pointed out that the original description of *Lilium*

Fig. 14.
1–5 *Lilium pumilum* Delile: 1 Upper part of plant; 2 Bulb; 3 Pistil; 4 Outer perianth segment; 5 Inner perianth segment.
6–9 *L. duchartrei* Franch.: 6 Upper part of plant; 7 Part of middle part of plant; 8 Outer perianth segment; 9 Inner perianth segment.

pumilum in Redouté's *Liliaceae* was not by De Candolle, as had been thought, but by another French botanist, Delile. I have therefore changed the authority for the name given in the *Flora RPS*.

30 (30). **Lilium cernuum** Komar. in *Act. Hort. Petrop.* 20: 461. 1901; Woodc. et Stearn, *Lil. World* 192, f. 34. 1950.

Bulb oblong or ovoid, 4cm tall; scales lanceolate or ovate, white. Stem *c.* 65cm tall, glabrous. Leaves very narrowly linear, 8–12cm long, 2–4mm wide, apex acuminate, margins slightly revolute and also papillose, midrib conspicuous. Raceme with 1–6 flowers; bracts leaf-like, linear, *c.* 2cm long, not thickened at the tip; pedicels 6–18cm long, erect, decurved near the tip; flowers nodding, fragrant; perianth segments lanceolate, revolute, 3.5–4.5cm long, 8–10mm wide, blunt at the apex, pale purple-red, with deep purple spots towards the base, nectaries densely papillose; filaments *c.* 2cm long, glabrous, anthers 1.4cm long, dark purple; ovary cylindric, 8–10mm long, 2mm broad; style 1.5–1.7cm long.

Flowering period: July.

Occurs in Jilin. It grows among other herbs or in thickets. It is also distributed in Korea and the USSR.

This species is similar to *L. pumilum* Delile, but it may be distinguished by its pale purple-red flowers with deep purple basal spots.

† *Lilium cernuum* has a much smaller range than its close relative *L. pumilum*, and is apparently nowhere very numerous. It occurs in Korea as far south as the Diamond Mountains (Kumgang San), and crosses the Changbai Shan range into China. It is also recorded from beyond the Ussuri in the USSR. It probably reached cultivation in western Europe via Petrograd, in about 1914, from collections made by its discoverer Komarov. It has proved reasonably easy to grow and has persisted ever since. It favours slightly cooler and moister conditions than *L. pumilum*, and usually grows rather taller and is less floriferous. The pink or rosy-purple coloration of the flowers makes this a very attractive addition to the colour range of this group of lilies. *L. cernuum* has been much used as a parent of hybrids, both in Britain and in the USA.

31 (31). **Lilium callosum** Sieb. et Zucc., *Fl. Jap.* 1: 86, t. 41. 1839. – *L. talanense* Hay., *Icon. Pl. Form.* 4: 98. 1914.

Bulb small, flattened-globose, 2cm tall, 1.5–2.5cm in diameter; scales ovate or ovate-lanceolate, 1.5–2cm long, 6–12mm wide, white. Stem 50–90cm tall, glabrous. Leaves scattered, linear, 6–10cm long, 3–5mm wide, 3-nerved, glabrous, margins papillose. Flowers solitary or rarely several in a raceme, nodding; bracts 1–2, 1–1.2cm long, thickened at the tip; pedicels 2–5cm long, decurved; perianth segments oblanceolate-spathulate, 3–4cm long,

4–6mm wide, revolute from the middle upwards, red or light red, almost unspotted, nectaries sparsely papillose; filaments 2–2.5cm long, glabrous, anthers 7mm long; ovary cylindric, 1–2cm long, 1–2mm broad; style shorter than ovary, stigma swollen, trilobed. Capsule narrowly oblong, *c.* 2.5cm long, 6–7mm broad.

Flowering period: July–August. Fruiting period: August–September.

Occurs in Taiwan, Guangdong, Zhejiang, Anhui, Jiangsu, Hubei, Henan and Manchuria. It grows on mountain slopes or among other herbs, at 182–640 metres above sea level. It is also distributed in Korea, Japan and the USSR.

This species is similar to *L. fargesii* Franch., but differs in having bracts which are thickened at the tip, more or less unspotted red or light red flowers, and nectaries which are papillose on each side.

† This little lily is also similar to *L. pumilum*, from which it may be most readily distinguished by its short style (shorter than the ovary) and by the thickened tips of its bracts (and often also of its perianth segments). Its flowers are also of a much duller shade of red. It has been in cultivation for over a century, and has been re-introduced more than once, but is not often seen in gardens now. It is, however, little harder to cultivate than *L. pumilum*, and requires similar conditions. As it often flowers as late as August, it ought to be worth growing to extend the flowering period of this group of lilies.

In the wild it has an extraordinary range of distribution, from Taiwan and Guangdong in the south along most of the east coast of China (though missing out Fujian and Shandong) as far north as Manchuria and neighbouring parts of the USSR. It is the only East Asian lily to have such an extensive north-south range, yet has a rather narrow east-west distribution, except for a westward extension to the Yichang area of Hubei. The reasons for this are not at all clear. Perhaps it was originally even more wide-ranging, and has died out in much of its former range, for it appears to be a rare plant in most of the areas where it currently occurs. It is most common in southern Japan. On Okinawa in the Ryukyu Islands a yellow variety, var. *flaviflorum* Makino, is found.

32 (28). **Lilium papilliferum** Franch. in *Journ. de Bot.* 6: 316. 1892; Woodc. et Stearn, *Lil. World* 293, f. 103. 1950.

Bulb ovoid, 3cm tall, 2.5cm in diameter; scales ovate or lanceolate-ovate, white. Stem *c.* 60cm tall, tinged with purple, densely papillose. Leaves numerous, scattered on the middle and upper parts of the stem, linear, apex acute, 5.5–7cm long, 2.5–4mm wide, midrib conspicuous. Raceme with *c.* 5 flowers; bracts leaf-like, 4–5cm long, 3–5cm wide; pedicels 4.5–5cm long; flowers fragrant, nodding, purple-red, perianth segments oblong, acute at the apex, slightly narrowed at the base, 3.5–3.8cm long, 1–1.3cm wide,

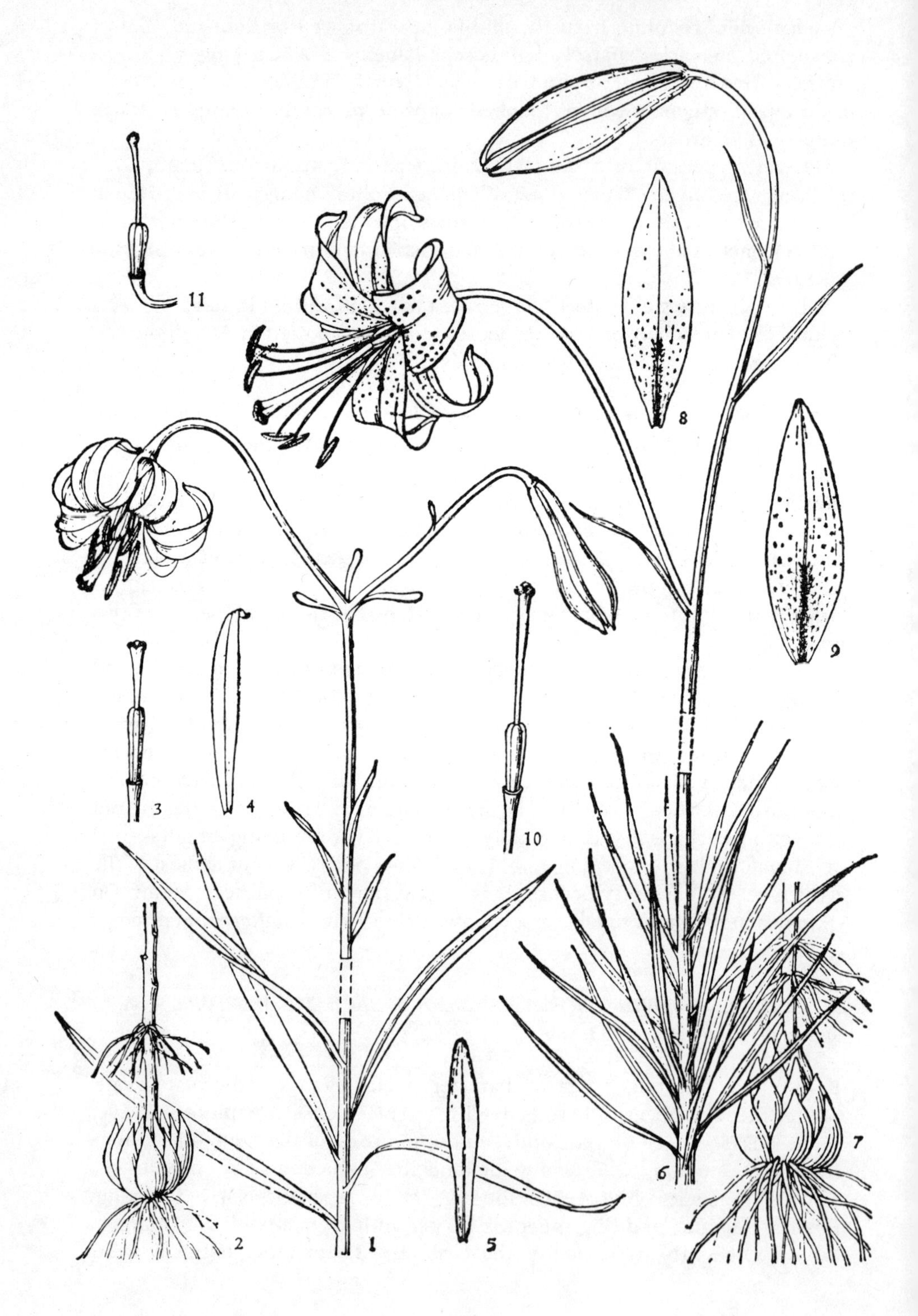
11
8
9
3
4
10
2
1
5
6
7

nectaries with papillae and cristate projections on each side; filaments 2cm long, glabrous, anthers light brown, pollen orange; ovary cylindric, 1cm long, 4mm broad, style 1.3cm long. Capsule oblong, 2–2.5cm long, 1.5–2cm broad.

Flowering period: July. Fruiting period: September.

Occurs in Yunnan (north-west part), Sichuan (western part) and Shaanxi (southern slopes of the Qinling Mountains). It grows on mountain slopes among shrubs, at 1000–3000 metres above sea level.

† This is a rare lily, both in cultivation and in the wild. It has been collected by westerners on only a few occasions, first by Père Delavay in 1888, and then not for 26 years until Forrest found it in 1914. It was subsequently collected a few more times by Rock as well as Forrest, always in north-west Yunnan. The distribution given in the *Flora RPS* is considerably larger than previously recorded. The stock now in cultivation probably all derives from bulbs and seeds collected by Dr Rock in the Lijiang mountains and sent to the USA and Britain in 1948.

L. papilliferum is hardy, and starts into growth very late, thus avoiding being damaged by frost. It often flowers in gardens as late as September. This is one reason for its scarcity in cultivation, for it hardly has time to ripen seed before the onset of winter. Seed germinates readily when it is available. The flowers are frequently very dark in colour, sometimes almost black, so that this is an attractive and unusual lily, and it is a pity that it has not proved more amenable to cultivation.

33 (32). **Lilium fargesii** Franch. in *Journ. de Bot.* 6: 317. 1892; Woodc. et Stearn, *Lil. World* 214, f. 45. 1950. – *L. cupreum* Lévl. in *Bull. Acad. Intern. Géogr. Bot.* 25: 38. 1915.

Bulb ovoid, 2cm tall, 1.5cm in diameter; scales lanceolate, 1.5–2cm long, *c.* 6mm wide, white. Stem 20–70cm tall, 2–4mm broad, papillose. Leaves scattered on the middle and upper parts of the stem, linear, 10–14cm long, 2.5–5mm wide, apex acuminate, margins revolute, glabrous on both surfaces. Flowers solitary or several in a raceme, nodding; bracts leaf-like, 2.3–2.5cm long, not thickened at the tip; pedicels 4–5.5cm long, slightly decurved near the tip; perianth greenish-white, densely spotted with

Fig. 15.
1–5 *Lilium callosum* Sieb. et Zucc.: 1 Upper part of plant; 2 Bulb; 3 Pistil; 4 Outer perianth segment; 5 Inner perianth segment.
6–10 *L. davidii* Duchartre: 6 Upper part of plant; 7 Bulb; 8 Outer perianth segment; 9 Inner perianth segment: 10 Pistil.
11 *L. cernuum* Komarov: Pistil.

4
7
9
8
1
2
5
6
3

purple-brown; perianth segments lanceolate, 3–3.5cm long, 7–10mm wide, revolute, nectaries with cristate projections on each side; filaments 2–2.2cm long, glabrous, anthers long-oblong, 7–9mm long, 2mm broad, orange; ovary cylindric, 1–1.5cm long, 2mm broad; style 1.2–1.5cm long, stigma slightly swollen, trilobed. Capsule oblong, 2cm long, 1.5cm broad.

Flowering period: July–August. Fruiting period: September–October.

Occurs in Yunnan, Sichuan, Hubei and Shaanxi. It grows in woodland on mountain slopes, at 1400–2250 metres above sea level.

This species is similar to *L. callosum* Sieb. et Zucc., but differs in having bracts not thickened at the tip, greenish-white flowers with purple spots, and cristate projections on each side of the nectaries.

† This species is uncommon in the wild and has never been in cultivation. It was consigned to obscurity by E. H. Wilson, who saw it on a few occasions but did not attempt to introduce it. 'As far as my memory serves', he wrote, 'it had a rather unpleasant odour. It is the least attractive of the Lilies of Eastern Asia, and when introduced will interest the collector only.' There are, I am sure, many collectors who would be pleased to be able to obtain this lily now! No lily is entirely unworthy of attention, if only for its potential for hybridization.

Though the *Flora RPS* says that this lily grows in woodland, Wilson knew it as 'native of the open, grassy country on the higher mountains of western Hupeh and contiguous Szech'uan'. It has been collected as far south as Dashuijing (Ta-ch'oui-tsin) south of Zhaotong in north-east Yunnan, and in the north extends into southern Shaanxi. If it ever were introduced, it would probably not be among the most difficult of lilies to grow. Its requirements are likely to be similar to those of *L. callosum*.

34 (33). **Lilium xanthellum** Wang et Tang, *Flora RPS* 14: 283. 1980.

Bulb large, sub-globose, 4.5cm tall, 4–5cm in diameter; scales lanceolate, 4–4.5cm long, 1.2–1.5cm wide, yellow. Stem 35–55cm tall, densely clothed with scales visible under a lens. Leaves scattered, linear, 4–8cm long, 2–3mm wide, margins slightly revolute and also papillose, midrib conspicuous. Flowers 1–2; bracts leaf-like, up to 3.5cm long, 2mm wide, not thickened at the tip; pedicels 3.5cm long; perianth segments yellow, unspotted, 3.5cm long, 6mm wide, apex blunt, nectaries with cristate projections on each side;

Fig. 16.
1–4 *Lilium fargesii* Franch.: 1 Upper part of plant; 2 Lower part of plant; 3 Inner perianth segment; 4 Outer perianth segment.
5–9 *L. xanthellum* Wang et Tang: 5 Upper part of plant; 6 Lower part of plant; 7 Inner perianth segment; 8 Stamen; 9 Pistil.

filaments 1.6–3cm long, glabrous; ovary 1.3–1.5cm long, 2–3mm broad; style 1.2–1.6cm long, stigma slightly swollen, trilobed.

Flowering period: June.

Occurs in Sichuan (Xiangcheng). It grows among shrubs on sunny mountain slopes, at 3200 metres above sea level.

This species is similar to *L. fargesii* Franch., but its bulb is large, 4.5cm tall, 4–5cm in diameter, and yellow, and the flowers are yellowish-green and unspotted.

var. **luteum** Liang, *Flora RPS* 14: 283. 1980.

This variety differs from the type in having purple-spotted flowers.

Occurs in Sichuan (Xiangcheng). It grows in gullies and at the bases of cliffs, at 3600 metres above sea level.

† This new species and its variety are entirely unknown in the west. The relationship to *L. fargesii* is obviously very close, and as flower-colour alone is not sufficient to merit the description of a new species, it is really only the size of the bulb which clearly distinguishes these two lilies. They are also geographically isolated, however, for Xiangcheng is in western Sichuan, close to the border with Yunnan and directly north of Zhongdian in that province. This location is some 250 miles distant from the nearest known station for *L. fargesii*. The description of *L. xanthellum* does not describe how the flowers are held, but the illustration shows them in a semi-erect position. The drawing was undoubtedly made from a herbarium specimen, however, so that this may well reflect the position they took on in the press, and should not be assumed to be their natural pose.

VIII. Section Martagon Duby

35 (37). **Lilium tsingtauense** Gilg in *Bot. Jahrb.* 34, Beibl. 75: 24. 1904; Woodc. et Stearn, *Lil. World* 359, f. 114. 1950.

Bulb sub-globose, 2.5–4cm tall, 2.5–4cm in diameter; scales lanceolate, 2–2.5cm long, 6–8mm wide, white, usually unjointed. Stem 40–85cm tall, not papillose. Leaves verticillate, in 1–2 whorls, each whorl consisting of 5–14 leaves, oblong-oblanceolate or oblanceolate to elliptic, 10–15cm long, 2–4cm wide, apex acute, base broadly cuneate, shortly petiolate, glabrous on both surfaces, with a few scattered leaves in addition to the verticillate leaves, the scattered leaves lanceolate, 7–9.5cm long, 1.6–2cm wide. Flowers solitary or 2–7 in a raceme, erect; bracts leaf-like, lanceolate, 4.5–5.5cm long, 0.8–1.5cm wide; pedicels 2–8.5cm long; perianth orange or vermilion, with purple-red spots; perianth segments long-elliptic, 4.8–5.2cm long,

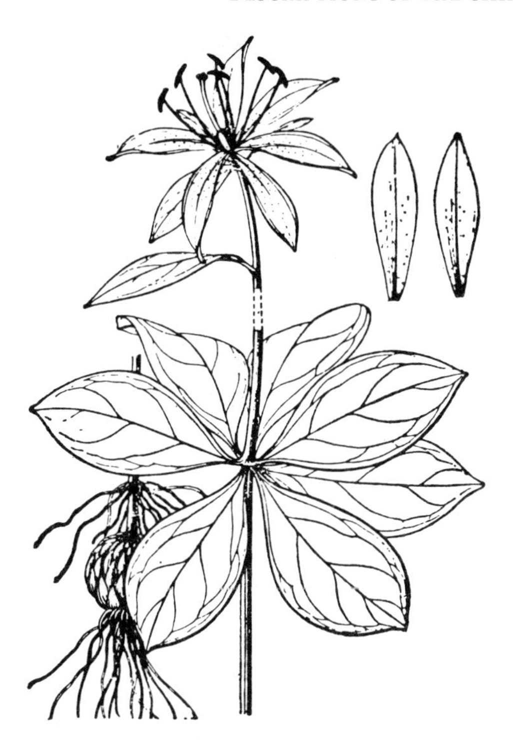

Fig. 17.
Lilium tsingtauense Gilg.

1.2–1.4cm wide, nectaries not papillose; filaments 3cm long, glabrous, anthers orange; ovary cylindric, 8–12mm long, 3–4mm broad; style twice as long as the ovary, stigma swollen, usually trilobed.

Flowering period: June. Fruiting period: August.

Occurs in Shandong and Anhui. It grows on sunny mountain slopes in mixed woodland or among tall herbs, at 100–400 metres above sea level. It is also distributed in Korea.

This species is very like *L. distichum* Nakai, but its bulb scales usually have no nodes, and its perianth segments are spread open and not revolute. Its area of distribution is also different.

† *Lilium tsingtauense* takes its name from the port of Qingdao (Tsingtau) on the coast of Shandong. It was first described from specimens collected in the Lao Shan hills east of the port during the period when Qingdao was a German concession. It is in this area that it is most common in China, though it is also recorded from Ya Shan, some 75 miles to the north-east, and from the vicinity of Weihai, on the north coast of the Shandong peninsular not far from its eastern tip. It also has an outlying station further to the south-west in Anhui province. It would seem now to be a rare lily in China. I have visited the Lao Shan range several times, but have never succeeded in finding this plant

(though it must be said that parts of the area are inaccessible, as there are military installations there). The range is of rather soluble limestone, reaching a maximum height of 1133 metres. *L. tsingtauense* is said to grow in moist places among open woodland or herbaceous growth on the lower slopes of the hills.

This lily is more widespread and apparently much more common in southern and central Korea, from Cheju Do (Quelpaert Is.) northwards to the Diamond Mountains. It has been grown in Japan since before 1841, and a yellow-flowered form is recorded in cultivation there. It was introduced to Britain before 1901, but did not at first seem to be a very strong-growing plant. It is nevertheless now well-established in western gardens, probably from post-war introductions from southern Korea, and today is considered to be quite easy to grow in a humus-rich soil in light shade. It is beginning to make a significant contribution to the development of new hybrids, for it crosses easily with other members of this section.

36 (39). **Lilium distichum** Nakai in Kamibayashi, *Chosen Yuri Dazukai*, t. 7. 1915; et in *Bot. Mag. Tokyo* 31: 6. 1917.

Bulb ovoid, 2.5–3cm tall, 3.5–4cm in diameter; scales lanceolate, 1.5–2cm long, 4–6mm wide, white, jointed. Stem 60–120cm tall, papillose. Leaves in a whorl of 7–9 (–20) on the middle of the stem, and also with a few scattered leaves, obovate-lanceolate to oblong-lanceolate, 8–15cm long, 2–4cm wide, apex acute or acuminate, gradually tapering towards the base, glabrous. Flowers 2–12 in a raceme; bracts leaf-like, 2–2.5cm long, 3–6mm wide; pedicels 6–8cm long; perianth pale vermilion, spotted with purple-red; perianth segments somewhat revolute, 3.5–4.5cm long, 6–13mm wide, nectaries not papillose; stamens shorter than the perianth segments, filaments *c.* 2–2.5cm long, glabrous, anthers linear, up to 1cm long; ovary cylindric, 8–9mm long, 2–3mm broad; style about twice as long as the ovary, stigma globose, trilobed. Capsule obovoid, 2cm long, 1.5cm broad.

Flowering period: July–August. Fruiting period: September.

Occurs in Jilin and Liaoning. It grows on mountain slopes among or at the edges of woodland, by the sides of roads or near water, at 200–1800 metres above sea level. It is also distributed in Korea and the USSR. The bulb contains starch and can be eaten or used to make wine.

This species is very like *L. tsingtauense* Gilg, but differs in having jointed bulb scales and revolute perianth segments. Its area of distribution is also different.

† This is an interesting species, which has been much confused with various closely related lilies in the past. Its taxonomic status is still not entirely resolved, and it seems that it may still be confused by Korean botanists with *L. tsingtauense* under the name of *L. miquelianum* Makino. It seems reasonably certain, however, that it is a distinct species, intermediate in form

Fig. 18.
Lilium distichum Nakai.

between *L. tsingtauense* and *L. medeoloides* A. Gray from Japan, southern Korea and the USSR. It is mainly distinguished by its flowers, which have only slightly revolute perianth segments and are often outward-facing. *L. tsingtauense* has erect flowers with segments that are not revolute, and *L. medeoloides* has nodding, more strongly revolute flowers.

L. distichum is quite common on both the Chinese and Korean sides of the Changbai Shan, and extends northwards in Jilin province to the banks of the Mudan Jiang. Southwards it occurs throughout northern and central Korea, including the Diamond Mountains and Mt Sorak. It seems to grow mainly in areas of volcanic rocks. It is rare in cultivation, but ought not to be too difficult to grow, if it behaves at all like its near relatives. Conditions similar to those for *L. tsingtauense* should suit it, though it might be expected to show a stronger preference for acid soils.

37 (38). **Lilium martagon** L., *Sp. Pl.* ed. 1, 303, 1753; Komar., *Fl. URSS*. 4: 288. 1935; Woodc. et Stearn, *Lil. World* 270. 1950.

Occurs in Europe, but not in China.

var. **pilosiusculum** Freyn in *Öst. Bot. Zeitschr.* 40: 224. 1890.

Bulb broadly ovoid, 3–5cm tall, 5cm in diameter; scales oblong, 2–2.5cm long, 8–10mm wide, apex acute, unjointed. Stem 45–90cm tall, streaked with purple, glabrous. Leaves verticillate, rarely scattered, lanceolate, 6.5–11cm long, 1–2cm wide. Flowers 2–7 in a raceme, nodding; bracts leaf-like, lanceolate, 2–4cm long, 5–6mm wide, apex acuminate, with white hairs on the margins and the undersurface and in the axils; pedicels decurved near the tip, 4.5–6cm long; perianth purple-red, spotted, with long, curly white hairs on the outside; perianth segments long-elliptic, 3.2–3.8cm long, 8–9mm wide, nectaries papillose; filaments 2.2–2.4cm long, anthers long-elliptic, 9mm long; ovary cylindric, 8–9mm long, 2–3mm broad, style 1.5cm long, stigma swollen. Capsule obovoid-oblong, 2–2.8cm long, 1.5–2cm broad, light brown.

Flowering period: June. Fruiting period: August.

Occurs in northern Xinjiang. It grows on shady mountain slopes or among shrubs under woodland, at 200–2500 metres above sea level. It is also distributed in Mongolia and the USSR. The bulb is edible.

† *Lilium martagon* is well-known to all with an interest in lilies. The Asian variety *pilosiusculum* is distinguished from the type mainly by its greater hairiness, particularly of its bracts and flowers, and by its narrower leaves. It is in cultivation in the west, but is not distinct enough from other forms of this species to cause much excitement. Over most of its wide range, from the Volga to the Lena, it is the only *Lilium* species to occur, and is the sole representative of its genus to reach Xinjiang. It is found there only in the Altai and Tarbagatai ranges, on the borders with Mongolia and the USSR. The discontinuity in distribution of the genus *Lilium* in north-west China suggests that it spread across Eurasia mainly along the Himalayas and through Afghanistan and Iran to the Caucasus. The close relationship between *L. polyphyllum* D. Don, which extends into eastern Afghanistan, and several of the Caucasian lilies (at least one of which is recorded from northern Iran) tends to confirm this.

Cardiocrinum (Endl.) Lindl.

Lindl., *Veg. Kingdom* 205. 1846. – *Lilium* sect. *Cardiocrinum* Endl., *Gen. Pl.* 141. 1836.

Bulb formed of the swollen bases of the petioles of the basal leaves, shrivelling immediately after flowering; bulblets numerous, ovoid, with fibrous tunics, without scales. Stem tall, glabrous. Leaves basal and cauline, the latter scattered, usually ovate-cordate, gradually diminishing in size up the stem, reticulately veined, petiolate. Inflorescence a raceme, with 3–16 flowers; perianth narrowly trumpet-shaped, white, streaked with purple; perianth segments 6, separate, more or less connivent, linear-oblanceolate; stamens 6, filaments compressed, anthers dorsally attached, versatile; ovary cylindric, style about twice as long as the ovary, stigma capitate, shallowly trilobed. Capsule oblong, with a small pointed projection at the tip, a short thick basal stalk, 6 blunt ridges and also numerous fine transverse striae. Seeds numerous, flattened, reddish-brown, narrowly winged around the circumference. Germination epigeal, often delayed.

Altogether 3 species, distributed in China, Japan and the Himalaya. China has 2 species, occuring in the Qinling Mountains and provinces to the south.

Key to the species of the genus *Cardiocrinum*.

1. Raceme of 10–16 flowers, without floral bracts; filaments about half the length of the perianth segments or a little longer. Plant robust, 1–2 metres tall, stem 2–3cm in diameter (Tibet, Sichuan, Yunnan, Shaanxi, Hubei, Hunan, Guangxi) .... 1. *C. giganteum* (Wall.) Makino
1. Raceme of 3–5 flowers, each flower with a single bract; filaments about two-thirds the length of the perianth segments. Plant somewhat smaller, 0.8–1 metre tall, stem 1–2cm in diameter (Hubei, Hunan, Jiangxi, Zhejiang, Anhui, Jiangsu)............ 2. *C. cathayanum* (Wilson) Stearn

1. **Cardiocrinum giganteum** (Wall.) Makino in *Bot. Mag. Tokyo* 27: 125. 1913. – *Lilium giganteum* Wall., *Tent. Fl. Nepal.* 21, tt. 12–13. 1824. – *L. giganteum* var. *yunnanense* Leichtlin ex Elwes in *Gard. Chron.* ser. 3, 60: 49, f. 18. 1916; Wilson, *Lil. East. As.* 96, t. 14. 1925 – *Cardiocrinum giganteum* var. *yunnanense* (Leichtlin ex Elwes) Stearn in *Gard. Chron.* ser. 3, 124: 4. 1948.

Bulblets ovoid, 3.5–4cm tall, 1.2–2cm in diameter, pale brown when dry. Stem erect, hollow, 1–2 metres tall, 2–3cm in diameter, glabrous. Leaves papery, reticulately veined; basal leaves ovate-cordate or more or less broadly oblong-cordate, stem leaves ovate-cordate, the lower ones 15–20cm long, gradually diminishing in size up the stem, those near the inflorescence boat-shaped. Raceme of 10–16 flowers, bractless; perianth narrowly trumpet-shaped, white, with pale purple-red streaks inside; perianth segments linear-oblanceolate, 12–15cm long, 1.5–2cm wide; stamens

6.5–7.5cm long, approximately half the length of the perianth segments, filaments gradually expanding towards the base, compressed, anthers long-oblong, *c.* 8mm long, *c.* 2mm broad; ovary cylindric, 2.5–3cm long, 4–5mm broad; style 5–6cm long, stigma swollen, shallowly trilobed. Capsule sub-globose, 3.5–4cm long, 3.5–4cm broad, with a small pointed projection at the tip and a short thick basal stalk, reddish-brown, with 6 blunt ridges and numerous fine transverse striae, 3-valved. Seeds flat, obtusely deltoid, reddish-brown, 4–5mm long, 2–3mm wide, with a pale reddish-brown, translucent, membraneous wing all round the circumference.

Flowering period: June–July. Fruiting period: September–October.

Occurs in Tibet, Sichuan, Yunnan, Shaanxi, Hubei, Hunan and Guangxi. It grows among other herbs beneath woodland, at 1450–2800 metres above sea level. It is also found in India, Nepal, Bhutan and Burma. The bulb is used medicinally.

† This is the most imposing of all the lilies, with its large leaves and tall stem holding its numerous white flowers at head height and above. It has been in British gardens since about 1847, and has proved moderately easy to cultivate in the woodland garden, planted quite shallowly in soil rich in leaf-mould and given as much moisture as possible in summer. It will tolerate a slightly limy soil. Late spring frosts will damage the young shoots if they are not adequately protected, but otherwise it is usually fairly trouble-free. The major problems with this lily are that it is too large for many gardens, and that it takes so long to grow to flowering size. Seed is normally produced in plentiful quantity, and if sown as soon as ripe will germinate the following spring (if kept too long it will become dormant, and may need two winters to break the dormancy). But it will then take about seven years for the bulbs to grow large enough to flower. After flowering, the main bulb dies, leaving a varying number of small bulblets. These may also be grown on, and should come to flowering more quickly than seedlings, though it is said that they produce an inferior inflorescence. Few gardeners would discard them on that account! This is certainly a lily which demands patience and persistence from those who grow it, and for many it will remain a plant to be gazed at in awe during visits to large public gardens.

The *Flora RPS* merges the two varieties of this lily that were formerly recognized. The differences between the Himalayan and west Chinese forms of *Cardiocrinum giganteum* are not very great, and they almost certainly intergrade. There does therefore seem to be little reason to maintain varietal distinction between them. Some authors have made much of the tendency of the Chinese form to open its flowers from the top downwards, but E. H. Wilson states that this is a variable characteristic, which is in any case occasionally found in Himalayan plants of this species. He believed that the chief distinguishing features were the dark stem and horizontally disposed flowers of the Chinese form. These features were certainly evident in plants of this lily which I have seen growing on Mt Emei in Sichuan. Further research is needed to confirm whether they are constant in *Cardiocrinum giganteum* from China.

This is a widespread species. Its eastern limits are in central China, where it occurs in western Hubei and Hunan, and in the west it reaches as far as Kashmir. It occurs as far north as the southern slopes of the Qinling range in Shaanxi, at about 33° 40′ N., and in southern Yunnan it extends to just within the tropics. It is very much a woodland plant, growing often in quite dense shade, though also in open glades. It is abundant in much of its distribution area.

Fig. 19.
Cardiocrinum cathayanum (Wilson) Stearn.

2. **Cardiocrinum cathayanum** (Wilson) Stearn in *Gard. Chron.* ser. 3, 124: 4. 1948. – *Lilium cathayanum* Wilson, *Lil. East. As.* 99. 1925; Hu et Chun, *Icon. Pl. Sin.* t. 49. 1927.

Bulblets 2.5cm tall, 1.2–1.5cm in diameter. Stem 50–150cm tall, 1–2cm in diameter. Stem naked for *c.* 25cm above the basal leaves, the few lowest cauline leaves clustered together, the remainder scattered; leaves papery, reticulately veined, ovate-cordate to ovate, apex acute, base more or less cordate, 10–22cm long, 6–16cm wide, deep green above, pale green below;

petioles 6–20cm long, broadened at the base. Raceme with 3–5 flowers; pedicels short and thick, inclined obliquely upwards, each flower with a single bract; bracts oblong, 4–5.5cm long, 1.5–1.8cm wide; perianth narrowly trumpet-shaped, milk-white or pale green, with purple streaks on the interior; perianth segments linear-oblanceolate, 13–15cm long, 1.5–2cm wide, those of the outer whorl acute at the apex, those of the inner whorl blunter; filaments 8–10cm long, two-thirds the length of the perianth segments, anthers 8–9mm long; ovary cylindric, 3–3.5cm long, 5–7mm broad; style 6–6.5cm long, stigma swollen, shallowly trilobed. Capsule subglobose, 4–5cm long, 3–3.5cm broad, reddish-brown. Seeds flattened, reddish-brown, with a membraneous wing all round the circumference.

Flowering period: July–August. Fruiting period: August–September.

Occurs in Hubei, Hunan, Jiangxi, Zhejiang, Anhui and Jiangsu. It grows on mountain slopes in damp shady places among woodland, at 600–1050 metres above sea level. The capsule is used medicinally.

† *Cardiocrinum cathayanum* is the least known, to westerners at least, of the three *Cardiocrinum* species. It was not described until 1925, and a certain amount of doubt as to its distinctness from the Japanese *C. cordatum* (Thunb.) Makino has persisted in the minds of several recent authorities on the genus. The Chinese plant has only been in cultivation once, as far as is known, and then only very briefly, and it is not well represented in British herbarium collections. On the occasion when it did appear in cultivation, however, it had been supplied by a Japanese nurseryman under the name of the Japanese species, and was sufficiently obviously distinct to be quickly recognized. Bulbs were received in the spring of 1939, flowered in August of the same year, and died without producing either viable seed or bulblets, probably because they were in poor condition after their long journey from the Far East. One of the plants was photographed in flower, and is illustrated in the Royal Horticultural Society's *Journal*, vol. 70 (1945), figs. 27–28. It was much smaller than is normal in the wild, with only one flower, but this might be expected in the circumstances. Its leaves are mostly clustered together on the stem at about a third of its height, more or less forming a whorl, with a couple of much smaller leaves growing alternately above. The persistent bracts which are a feature of this species are very obvious in the plant photographed, and so also is the shape of the leaves, which are noticeably longer than broad and reniform-cordate at the base. In *Cardiocrinum cordatum* the leaves are as wide as long, and are deeply cordate. There seems no reason at present to question the separation of these two species.

In the wild *C. cathayanum* is moderately widespread, occurring from western Hubei and Hunan through all the four provinces eastwards to the sea. But it is not very common in any part of its range, and is confined to rather scattered localities where there is suitable woodland habitat on mountain slopes. Near the east coast, it is recorded from Tian Tai Shan in Zhejiang, and northwards to hills near Yixing in southern Jiangsu. It has been collected in the Lu Shan range in Jiangxi, and around Liantuo (Nanto) and Changyang in Hubei.

If it ever reaches western gardens again, it would require similar conditions to other members of the genus. It has been said to be the least desirable of the *Cardiocrinum* species, but its lesser stature ought to make it more suitable for smaller gardens than *C. giganteum*, and it should certainly be sufficiently decorative to be grown for its merits and not merely for interest.

Nomocharis Franch.

Franch. in *Journ. de Bot.* 3: 113. 1889.

Bulb ovoid to ovoid-globose, composed of a number of scales, white, brown when dry. Stem 25–100 (–150)cm tall, glabrous or papillose. Leaves scattered or verticillate, lanceolate, ovate-lanceolate or elliptic-lanceolate. Flowers solitary or several arranged in a raceme, open, pink, red, white or pale yellow; perianth segments 6, separate, those of the outer whorl usually narrower, finely spotted or blotched, entire, those of the inner whorl broader, usually blotched or spotted, entire or with fringed or irregularly serrate margins, with fleshy, purple-red, cushion-shaped swellings on the inside at the base; stamens 6, filaments expanded in the lower part into a fleshy cylinder or not so expanded, filiform in the upper part; anthers ellipsoid, dorsally attached, versatile; ovary cylindric, style gradually expanded towards the tip, stigma capitate, shallowly trilobed. Capsule oblong-ovoid, brown.

Six or seven species, distributed in south-west China and adjacent areas of Burma and India. China has six species, occurring in Yunnan, Sichuan and Tibet. This genus is very similar to the genus *Lilium*. Those species with fleshy, cushioned-shaped swellings at the base of the inner perianth segments are included in this genus, all those without such swellings are placed in the genus *Lilium*.

Key to the species of the genus *Nomocharis*

1. Leaves all scattered; filaments sub-filiform, slightly expanded in the lower part, compressed, but definitely not enlarged into a fleshy cylinder.
   2. Style as long as or a little shorter than the ovary; perianth segments finely spotted near the base (Yunnan, Sichuan, Tibet) .... 1. *N. saluenensis* Balf. f.
   2. Style up to twice as long as the ovary; perianth segments with at least a few spots near the base, or spotted and blotched over much of, or the whole surface (Yunnan, Sichuan) ...... 2. *N. aperta* (Franch.) Wilson
1. Leaves verticillate or simultaneously verticillate, opposite and scattered on the same plant; filaments expanded in the lower part into a fleshy cylinder.
   3. Inner perianth segments spotted or blotched, outer perianth segments spotted or unspotted.
      4. Leaves narrowly to broadly elliptic or narrowly to broadly lanceolate, rarely ovate; perianth spotted and blotched all over, or spotted for a short distance near the base; inner perianth segments broadly ovate to sub-orbicular, more or less as broad as long, margins erose-lacerate or minutely erose (Yunnan, Sichuan) .... 3. *N. pardanthina* Franch.

4. Leaves linear to narrowly elliptic or lanceolate; inner perianth segments ovate to broadly elliptic, longer than broad, margins entire or minutely erose or slightly lacerate.
    5. Perianth spotted or blotched for a short distance near the base (Yunnan) .... 4. *N. farreri* (W. E. Evans) Harrow
    5. Perianth blotched all over (Yunnan, Tibet) ....5. *N. meleagrina* Franch.
3. Both inner and outer perianth segments unspotted (Yunnan) ....6. *N. basilissa* Farrer ex W. E. Evans

1 (*Lilium* 36). **Nomocharis saluenensis** Balf. f. in *Trans. Bot. Soc. Edinb.* 27: 294. 1918; Stapf in *Bot. Mag.* t. 9296. 1933. – *Lilium apertum* var. *thibeticum* Franch. in *Journ. de Bot.* 12: 221. 1898. – *Lilium saluenense* (Balf. f.) Liang, *Flora RPS* 14: 154. 1980.

Bulb ovoid, 2–4cm tall, 2–2.5cm in diameter, white. Stem 30–90cm tall, glabrous. Leaves scattered, lanceolate, 3.5–7cm long, 0.8–1.5cm wide. Flowers 1–7, opening wide into a saucer-shape, pink, with tiny purple spots on the inside near the base; outer perianth segments elliptic to narrowly elliptic, 3.5–5.2cm long, 1.6–2cm wide, apex acute, margins entire; inner perianth segments similar to the outer ones, 3–4cm long, 1.7–2cm wide, apex acute, with conspicuous tiny basal spots, margins entire; filaments subulate, *c.* 1cm long, anthers 3–4mm long; ovary 6–7mm long, 2.5–3mm in diameter; style shorter than the ovary, 2.5–4mm long, gradually expanded towards the tip, stigma capitate, shallowly trilobed. Capsule oblong, 1.7–1.8cm long, *c.* 1.8cm in diameter, purplish-green to brown.

Flowering period: June–August. Fruiting period: August–September.

Occurs in Yunnan (north-western part), Sichuan and Tibet (south-eastern part). It grows on mountain slopes in forest, at the edges of woods and in meadows, at 2800–4500 metres above sea level. It is also found in Burma.

† *Nomocharis* generally are not particularly easy to grow, though this is one of the easier species. The cultivation requirements of all the species are similar, being for a well-drained soil which must never dry out during the growing season but should be as dry as possible in winter. They prefer a cool climate, and have been most successfully grown in central Scotland, though there has also been some success in flowering them in southern England, Canada and the USA. The first recorded *Nomocharis* to bloom in cultivation was *N. pardanthina* at the Royal Botanic Garden, Edinburgh, in 1914, from seed collected by George Forrest. Most of the other species were introduced during the next two decades, and have persisted ever since.

Seed is usually plentifully produced by *Nomocharis* in cultivation, and is the best method of propagation. Scaling may also be used, as with *Lilium* species, but the bulbs of *Nomocharis* are rather small, and as they strongly resent root disturbance, lifting established bulbs to remove scales is a risky undertaking. Seed sown in late autumn or early spring will usually germinate as soon as there is sufficient warmth to permit growth. It is best to sow thinly, so that the seedlings may be grown on in their pots for a couple of seasons without any disturbance. They may then be transferred into large pots or boxes to grow on for another two or three years, before being planted into their final garden site when more or less at flowering size. When transplanting, it is best to turn out the whole pot or box and not separate the individual bulbs, to minimize damage to roots. In southern parts of Britain a shady situation should be chosen, though this is not necessary in cooler areas further north. Ground cover is not always necessary, but several growers have found it helpful to plant among dwarf shrubs or prostrate alpines such as *Cyananthus* and Sino-Himalayan gentians. A well-drained site in a peat garden is most likely to suit these beautiful lilies.

N. saluenensis was first collected by Soulié in 1895, on a mountain above Cigu (Tseku), near Dêqên in extreme north-west Yunnan. It was later found several times by George Forrest, and cultivated plants probably derive almost entirely from his collections. It flowered at the Royal Botanic Garden, Edinburgh, as early as 1927. Its wild range is very restricted, from a southern limit of about 27° N. in north-west Yunnan, to as far north as 28° 40′ in south-east Tibet. In the west it occurs on the border ranges between Yunnan and Burma, and just extends into Sichuan in the east. It has a greater altitudinal range than the other species of the genus.

There is considerable variation in the colour of the flowers of this species, from white, more or less spotted with red, through pink to rose-purple, sometimes with some pale yellow coloration on the inside of the perianth. One colour form was at one time considered to be a distinct species, *N. tricolor* Balf. f., but this cannot be upheld. The usefulness of recognizing colour-forms is questionable, as variation is continuous and all kinds of intermediates occur.

The cultivated plant formerly considered to be a 'stoloniferous' form of this species has now been described by J. R. Sealy as a hybrid, *N.* × *notabilis*, probably between *N. saluenensis* and *N. farreri*.

2 (*Lilium* 35). **Nomocharis aperta** (Franch.) Wilson, *Lil. East. As.* 13. 1925. – *Lilium apertum* Franch. in *Journ. de Bot.* 12: 220. 1898, excl. var. *thibeticum*; Liang, *Flora RPS* 14: 152, 1980.

Bulb ovoid, 1.5–2.5cm tall, 1–2cm in diameter; scales ovate-lanceolate,

Fig. 20.
1–6 *Nomocharis aperta* (Franch.) Wilson: 1–2 Whole plant; 3 Outer perianth segment; 4 Inner perianth segment; 5 Stamen; 6 Pistil.
7–12 *N. saluenensis* Balf. f.: 7 Bulb; 8 Upper and middle parts of plant; 9 Outer perianth segment; 10 Inner perianth segment; 11 Stamen; 12 Pistil.

4
3
11
12
10
1
2
5
6
8
9
7

yellowish-brown when dried. Stem 25–50cm tall, glabrous, approximately the lower quarter leafless. Leaves scattered, broadly to narrowly lanceolate, 3–5.5cm long, 0.8–1.2cm wide. Flowers 1–2, rarely 4, opening widely into a saucer shape, red, pink or pale yellow; outer perianth segments narrowly elliptic-lanceolate, 2.2–4.5cm long, 1.2–1.5cm wide, entire, with 3–8 purple-brown spots at the base; inner perianth segments ovate to broadly ovate, 2.2–4.3cm long, 1.3–1.6cm wide, apex acute, with a few to a dozen or more purple-red spots near the base; filaments subulate, *c.* 1cm long; ovary 5–7mm long, 2–2.5mm in diameter; style gradually expanded towards the tip, about twice as long as the ovary, 1–1.2cm long, stigma capitate, shallowly trilobed. Capsule oblong, *c.* 1cm long, 1.2cm broad, pale brown.

Flowering period: June–July. Fruiting period: September–October.

Occurs in Yunnan (north-west part). It grows in mixed woodland or meadows on mountain slopes, at about 3500 metres above sea level.

This species is very similar to *N. saluenensis* Balf. f., but differs in having its style longer than the ovary, and in having more heavily spotted perianth segments.

† In the *Flora RPS* this species, as described above, is placed in the genus *Lilium*, section Sinomartagon. This does not seem to be justifiable. The Chinese *Flora* also separates this species from *Nomocharis forrestii* Balf. f., a treatment not upheld by western botanists since 1925. The description of *N. forrestii* is as follows:

2a (1). **Nomocharis forrestii** Balf. f. in *Trans. Bot. Soc. Edinb.* 27: 293. 1918; Liang, *Flora RPS* 14: 160. 1980.

Bulb ovoid, 2.5–3.5cm tall, 2–2.5cm in diameter, yellowish-white. Stem 30–100 (–150)cm tall, glabrous. Leaves scattered, lanceolate or ovate-lanceolate, (2–) 2.5–6cm long, 0.7–1.5cm wide, apex acuminate. Flowers 1–6; perianth open, saucer-shaped, pink to red, finely spotted on the inside at the base, the spots gradually increasing in size towards the upper part to become purple-red blotches; outer perianth segments ovate to elliptic, apex acute, 2.5–4.2cm long, 1.5–1.8cm wide, margins entire; inner perianth segments broadly elliptic, 2.5–4cm long, 1.7–2.2cm wide, apex acute, with two purple-red cushion-shaped swellings on the inside at the base; filaments 7mm long, slightly expanded in the lower part, compressed, but not

Fig. 21.
1–6 *Nomocharis aperta* (Franch.) Wilson (*N. forrestii* Balf. f.): 1–2 Whole plant; 3 Outer perianth segment; 4 Inner perianth segment; 5 Stamen; 6 Pistil.
7–12 *N. pardanthina* Franch. f. *punctulata* Sealy: 7–8 Whole plant; 9 Outer perianth segment; 10 Inner perianth segment; 11 Stamen; 12 Pistil.

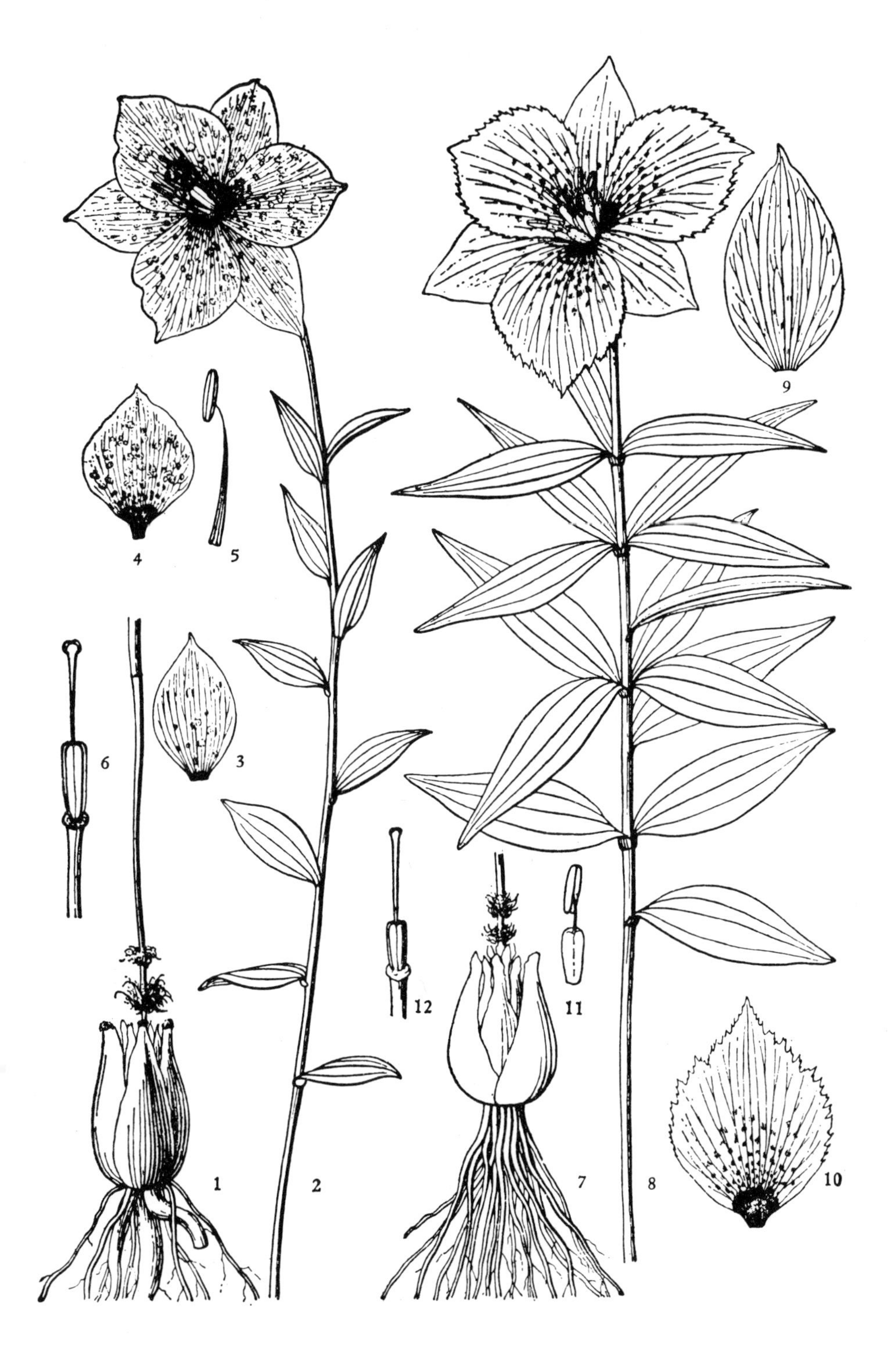
1
2
3
4
5
6
7
8
9
10
11
12

expanded into a fleshy cylinder, purple-red, upper quarter becoming fine, yellowish-white; ovary 7–9mm long, 2.5–3mm broad; style gradually expanded towards the tip, 6.5–8mm long, stigma capitate, shallowly trilobed. Capsule oblong-ovoid, 2.5cm long, 2cm broad, greenish-brown.

Flowering period: June–July. Fruiting period: August–October.

Occurs in Yunnan (north-west part) and Sichuan (south-west part). It grows on mountain slopes in woodland or in meadows, at 3000–3850 metres above sea level. It also occurs in Burma.

† *N. aperta* in the broad sense is a very variable species, and there is certainly continuous variation between the less robust form with lightly spotted flowers and indistinct nectarial swellings, described in the *Flora RPS* as *Lilium apertum*, and the generally larger form with heavily spotted and blotched flowers with rounded swellings on each side of the nectarial furrow, which the Chinese *Flora* accords distinct status as *Nomocharis forrestii*. A flower on a specimen of *N. aperta* collected near Weixi (Wei-Hsi) by Joseph Rock (no. 17158), for example, has no swelling of any kind on one inner segment, with fleshy ridges or flaps on the other two. The maculation also varies considerably from specimen to specimen. McLaren D247 from the Fu Chuan Shan, west of Weixi, is particularly unusual, having large purple-red blotches on the lower half only of the perianth segments, which merge into a more or less continuous band of colour. Moreover, some specimens which I have examined have filaments expanded in the lower part into a narrow cylinder, similar to the condition seen in *Nomocharis pardanthina* and other species, yet in all other respects are closest to *N. aperta*. One flower of a specimen collected by Kingdon Ward in the Adung valley on the Burma-Tibet frontier (no. 9551) has one expanded filament, while all the rest are of typical form. Another specimen (McLaren B45) from the Dali area has only very slight nectarial swellings but obviously swollen, narrowly cylindrical filaments (on two of the plants on the sheet). This great range of variation brings *N. aperta* very close not only to *N. saluenensis* at one extreme, but also to *N. meleagrina* at the other. It is quite impossible to maintain a distinction between *N. aperta* and *N. forrestii*, and it is also in my opinion very likely that *N. synaptica* Sealy from near the Tibetan border in Assam should be included in this species, though at present there are too few specimens of *N. synaptica* to be sure of this.

The total range of distribution of *N. aperta* includes the Muli area of Sichuan and all of north-west Yunnan as far south as the Cang Shan range above Dali. It may just extend into south-east Tibet, as it is recorded from mountains on the Yunnan-Tibet border and from the Burma-Tibet frontier. In the north of Burma it reaches as far west as 97° 13′ E. (which brings it within seventy miles of the recorded locations of *N. synaptica*).

In cultivation this has proved to be the most adaptable species of *Nomocharis*, and has probably survived in more gardens than any of the others. It is rarely as tall and robust as *N. pardanthina*, however. It was introduced to Britain by Forrest, and first flowered in Scotland in about 1929. It has also been grown successfully in Vancouver and Oregon in North America, and in New Zealand. It was awarded an AM in 1936.

3 (4). **Nomocharis pardanthina** Franch. in *Journ. de Bot.* 3: 113, t. 3. 1889. – *N. mairei* Lévl. in *Rep. Sp. Nov. Fedde* 12: 287. 1913; Liang, *Flora RPS* 14: 161. 1980.

Bulb sub-ovoid, 2.5–3.5cm tall, 2–3.5cm in diameter, white, pale brown when dried. Stem 25–65cm tall, papillose. Leaves verticillate, 4–8 in each whorl, lanceolate, 5–7cm long, 1–1.4cm wide, deep green above, pale green below, apex acuminate, both surfaces glabrous. Flowers one to several, nodding, white or pink; outer perianth segments ovate, 2.5–3.5cm long, 1.5–2cm wide, apex acuminate, margins entire; inner perianth segments broadly ovate to sub-orbicular, 2–3cm long, 2–3cm wide, margins erose-lacerate, densely covered with purple-red spots, the spots gradually increasing in size towards the upper part to become blotches, apex acute, base with purple-red, fleshy, cushion-shaped, cristate swellings; filaments expanded in the lower part into fleshy cylinders, 6.5–7mm long, filiform in the upper part, 2mm long, yellowish-white; ovary 6–8mm long, 2–3mm broad; style gradually expanded towards the tip, 6–8mm long, stigma capitate, shallowly trilobed. Capsule oblong-ovoid, 2.5cm long, 2.5cm broad, bluntly hexangular, pale brown.

Flowering period: May–July. Fruiting period: July–August.

Occurs in Yunnan and Sichuan. It grows on mountainsides at the edges of woods or on grassy slopes, at 2700–4050 metres above sea level.

This species is very like *N. meleagrina* Franch., but differs in having sub-orbicular perianth segments which are as broad as long.

† According to J. R. Sealy's revision of the genus *Nomocharis* published in 1983, the plant long grown in the west as *Nomocharis mairei* should in fact be referred to *N. pardanthina* Franch., the type specimen of which has innner perianth segments blotched all over and with erose-lacerate margins. *N. pardanthina* of gardens he considers to be so close as to be placed in the same species, but since it differs in having much less spotted flowers with the margins of the inner segments almost entire or only shallowly toothed, he accords it separate status as a form, *f. punctulata*. The *Flora RPS* was published before his revision appeared, and follows the previously-accepted treatment. As it seems to me that Sealy's conclusions are correct, I have amended the Chinese account accordingly.

3a (3). **N. pardanthina** Franch. f. **punctulata** Sealy in *Notes Bot. Gard. Edinb.* 36: 295. 1978. – *N. pardanthina* sec. *Gard. Chron.* ser. 3, 59: 314, f. 138. 1916; Woodc. et Stearn, *Lil. World*, 384, 388. 1950; Liang, *Flora RPS* 14: 161. 1980.

Bulb ovoid-globose, 2.5–3cm tall, 2.3–2.5cm in diameter, brown when dried, outer layer tinged with yellow. Stem 25–90cm tall, glabrous. Leaves

5
6
11
12
9
10
3
2
8
7
1
4

simultaneously scattered and verticillate on the same plant, narrowly elliptic or lanceolate-elliptic, 2.5–4.5cm long, 0.7–1.5cm wide. Flowers solitary or rarely several, red or pink; outer perianth segments ovate, 2.5–3cm long, 1.2–1.5cm wide, almost unspotted, margins entire; inner perianth segments broadly ovate to ovate-orbicular, 2.5–3cm long, 2–2.5cm wide, with purple-red spots and purple-red cushion-shaped swellings at the base on the inside, margins irregularly serrate; filaments expanded in the lower part into fleshy cylinders, purple-red or pink, 5–7mm long, filiform in the upper part, white, 2–2.5mm long; ovary 5–7mm long, 2–4mm broad; style gradually expanded towards the upper part, *c.* 8mm long, stigma capitate, shallowly trilobed. Capsule oblong, 2.5cm long, 1.7cm broad.

Flowering period: May–June. Fruiting period: July.

Occurs in north-west Yunnan. It grows on grassy slopes, at 3000–3500 metres above sea level.

† *Nomocharis pardanthina* was the first species of the genus to be discovered, being collected by Delavay in the Cang Shan range near Dali in 1883. It became the type of the new genus *Nomocharis,* described by Franchet in 1889. Subsequently it was collected again on several occasions from the Dali area; several plants of the type variety are now growing in the peat garden of the Royal Botanic Garden, Edinburgh, from bulbs collected in the Cang Shan by the joint Sino-British Expedition of 1981. It has also been found on the mountains to the north, west of Eryuan and Jianchuan and in the area around Weixi, in the Lijiang range and north-east of Zhongdian, and then eastwards through the area between Yongning and Yongsheng (Yungpeh) to the Muli region of Sichuan. Further east it occurs in the Da Liang Shan about 30 miles beyond Xichang, and in the neighbouring mountains of north-west Yunnan, where Maire collected the type specimen of *N. mairei.* Within this range, f. *punctulata* occurs in the Lijiang mountains, near Songgui (Sungkwe), and west of Jianchuan. There is also one specimen from much further west on the Longchuan Jiang-Nu Jiang (Shweli-Salween) divide.

This is a very beautiful plant in both its forms, and both have been awarded an FCC. They came into cultivation at the same time, first flowering in 1914 at the Royal Botanic Garden, Edinburgh, and have persisted ever since. Only in a few gardens have they really flourished, however. Occasionally some plants have been known to produce bulbils on the stem,

Fig. 22.
1–6 *Nomocharis pardanthina* Franch. (*N. mairei* Lévl.): 1–2 Whole plant; 3 Outer perianth segment; 4 Inner perianth segment; 5 Stamen; 6 Pistil.
7–12 *N. meleagrina* Franch.: 7 Bulb; 8 Middle and upper parts of plant; 9 Outer perianth segment; 10 Inner perianth segment; 11 Stamen; 12 Pistil.

but this is not a usual characteristic. In cultivation both forms may sometimes reach 1.5 metres or more in height, but are normally much shorter. They have both contributed to the group of hybrids (probably with *N. farreri*) named *Nomocharis* × *finlayorum*, which are a beautiful race with a slightly stronger constitution than the species.

4 (3a). **Nomocharis farreri** (W. E. Evans) Harrow in *New Flora and Silva* 1: 76, f. 23. 1928; [Cox in *Country Life* 55: 66, 141, f. p. 65, 140. 1924, nomen subnudum.] – *N. pardanthina* var. *farreri* (Cox) Evans in *Notes Bot. Gard. Edinb.* 15: 20, 22, 24, 26. 1925; Liang, *Flora RPS* 14: 161. 1980.

Bulb ovoid or sub-globose, 2.5–3.5cm tall, 1.5–4cm in diameter; scales 2–2.5cm long, 6–12mm wide, yellowish. Stem 35–90cm tall. Leaves verticillate, linear to lanceolate, 3.5–11cm long, 2–20mm wide, apex long-acuminate, dark green above, paler below. Flowers solitary or several in a loose raceme, nodding or horizontal, white to pale pink or rose; outer perianth segments elliptic or broadly lanceolate to ovate, 2.5–5cm long, 1–2.5cm wide, with a few crimson or purple blotches and spots at the base, margins entire; inner perianth segments ovate to broadly ovate, 3–5cm long, 1.5–3.5cm wide, with purple-red spots for about 10mm from the base and dark purple-red cushion-shaped basal swellings, margins entire or shallowly erose; filaments expanded in the lower part into a fleshy cylinder, deep rose or purplish, 4–8mm long, filiform in the upper part, yellowish, 2–2.5mm long; ovary sub-cylindric, 6–10mm long, 1.5–3mm in diameter; style gradually expanded towards the upper part, stigma capitate, shallowly trilobed. Capsule broadly cylindric, 1.5–2.5cm long, 1.5–2cm broad.

Flowering period: May–June. Fruiting period: July–August.

Occurs in Yunnan (western part). It grows among other herbs on mountain slopes, at *c.* 2800 metres above sea level. It is also distributed in Burma.

† The above description is based to a large extent on that given by Sealy in his revision of the genus. The *Flora RPS* treats this species as a variety of *N. pardanthina*, saying only that it 'differs from the type in that its leaves are longer and narrower . . .'. Its flowers are in fact very similar in coloration and maculation to those of *N. pardanthina* f. *punctulata*, which is why these two species were formerly asociated together, but its long, narrow leaves bring it closer to *N. meleagrina* and *N. basilissa*. It also resembles them in having comparatively narrow perianth segments. All three species occur on the mountain ranges of the Yunnan-Burma border, with *N. meleagrina* being found in the north, *N. farreri* in the south, and *N. basilissa* in between. All three are separated in distribution from *N. pardanthina*, which grows on the ranges to the east. In this region where the mountain chains run from north to south, and are separated by deep river valleys, this pattern of distribution is significant in assessing relationships between species.

N. farreri grows on the Pianma (Hpimaw) Pass, and northwards along the mountains of the Nu Jiang-N'Mai Hka divide to about 26° 20′ N. It has also been collected further west on mountains in Burma. It was introduced to cultivation by Farrer in 1919, and again by Forrest a few years later, and has proved slightly less easy to grow than *N. pardanthina*.

5 (5). **Nomocharis meleagrina** Franch. in *Journ. de Bot.* 12: 196. 1898.

Bulb ovoid, white, *c.* 2.5cm tall. 2–2.8cm in diameter. Stem 35–100cm tall, papillose or rarely smooth. Leaves verticillate, 5–8 in each whorl, narrowly lanceolate to elliptic-lanceolate, 4.5–11cm long, 0.8–2 (–3.5)cm wide, apex acuminate, glabrous on both surfaces, sometimes with conspicuously papillose margins. Flowers 2–4 in a raceme, white or pink, nodding; outer perianth segments elliptic to ovate-elliptic, 4–5cm long, 1.8–2.5cm wide, apex acute, with purple-red blotches, margins entire; inner perianth segments ovate or broadly elliptic, 4–5cm long, 2.5–3cm wide, evenly covered with purple-red spots at the base, the spots gradually enlarging into blotches towards the upper part, margins irregularly serrate, apex acute, with deep reddish-brown, fleshy, cristate, cushion-shaped swellings at the base; filaments expanded in the lower part into fleshy cylinders, 6–7mm long, purplish-brown, in the upper part filiform, 2–2.5mm long, yellowish-white, anthers long-ellipsoid, 3–3.5mm long, *c.* 1mm broad; ovary 7–8mm long, *c.* 2mm broad; style gradually expanded towards the tip, 7–9mm long, stigma capitate, shallowly trilobed. Capsule oblong-ovoid, 2–2.5cm long, 2–2.5cm broad, pale brown.

Flowering period: June–July. Fruiting period: August–September.

Occurs in Yunnan (north-west part) and Tibet (south-east part). It grows on mountainsides in mixed woodland or at the edges of woods, or in meadows, at 2800–4000 metres above sea level.

This species is very like *N. pardanthina* Franch., but it has perianth segments which are elliptic, longer than broad.

† *N. meleagrina* is quite a distinct species, its heavily spotted flowers and whorled leaves readily distinguishing it from all other *Nomocharis* except *N. pardanthina*. It is generally a more slender plant than the latter species, with longer and relatively narrower leaves, and differs most noticeably in having inner perianth segments longer than wide and much less toothed at the margins. Although the amount of spotting on the flower differentiates it from *N. basilissa* and *N. farreri*, it resembles these two species in having long, narrow leaves, and is probably closely related to them. It occurs in the extreme north-west of Yunnan, from a southern limit in the Fu Chuan mountains near Weixi northwards along the mountains of the Lancang Jiang-Nu Jiang (Mekong-Salween) divide into Tibet. Its northern limit is at about 28° 40′, and it extends westwards to the mountains of the Nu Jiang-Drung Jiang (Salween-Kiu Chiang) divide. It is thus one of the most northerly species of the genus.

This is a very rare species in cultivation. It flowered in Britain in the 1930s, but was thought to have been lost subsequently, and was not rediscovered until 1968. It now survives in only a few gardens.

6 (2). **Nomocharis basilissa** Farrer ex W. E. Evans in *Notes Bot. Gard. Edinb.* 15: 22, t. 205, 212a. 1925.

Bulb small, ovoid; scales loose, lanceolate or ovate-lanceolate. Stem 35–95cm tall. Leaves simultaneously scattered and verticillate on the same plant, lanceolate or narrowly lanceolate, 5.5–9cm long, 5–7mm wide, apex long-acuminate, dark green above, bluish-grey below. Flowers solitary or 2–5 arranged in a loose raceme, nodding, red or white tinged with purple at the base; outer perianth segments elliptic-lanceolate or ovate-lanceolate, *c.* 4cm long, 1.6–2cm wide; inner perianth segments wider than the outer ones, 2–2.5cm wide, at the base with two deep purple cushion-shaped swellings cristate in the upper part; filaments expanded in the lower part into fleshy cylinders, purple, upper part fine, anthers *c.* 3mm long. Ovary 6–7mm long, style gradually expanded towards the tip, 8–10mm long, stigma large, trilobed.

Occurs in north-west Yunnan. It grows in alpine regions among thickets of dwarf bamboo or in meadows, at 3928–4255 metres above sea level. It is also distributed in Burma.

† This is the only Chinese *Nomocharis* species never to have been in cultivation. It has probably only ever been seen by two westerners, Reginald Farrer and George Forrest. It was Farrer who first saw it, and his form is the one which gardeners most regret, for it has flowers of a very striking colour, '. . . a pure luminous salmon scarlet, unspotted, like nothing so much as some wonderful strain of *Papaver orientale*', according to Farrer's field-note. He collected this plant twice, once on the Chawchi Pass in north-east Upper Burma, and once on the Mokuji Pass a few miles further north, on the Burma-Yunnan border. He noted that 'it is never an abundant plant . . . , but there is quite a lot of it, dotted sporadically and singly in the light cane-brakes, where, down on the Chinese side, it occasionally meets *Farrer 1785* [*Nomocharis saluenensis*]. When well-developed it can attain 3–4 feet, and can carry as many as five or six flowers, always absolutely pendant, not horizontal.' It was in the summer of 1920 that Farrer found this superb lily, and in October of that year he died while still in the field. His pressed specimens were sent back to Britain, but all his seed harvest of the season was lost. No other collector has ever been able to return to the area and bring this scarlet *Nomocharis* into cultivation.

The form which George Forrest collected further south on the N'Mai Hka-Nu Jiang (Salween) divide was less striking, having basically white flowers flushed with pink or purple-crimson. But it otherwise closely resembles Farrer's plant, having unspotted flowers and narrow leaves, and must be assigned to the same species.

The total known area of distribution of *N. basilissa* thus extends for about 90 miles from north to south along the range of mountains on the Burma-Yunnan border which separates the Nu Jiang from the N'Mai Hka. This is an area of heavy monsoon rainfall, with almost constant mist or rain from June to October, and it is probable that the scarlet *Nomocharis* would not be one of the more easily cultivated species of the genus. Unfortunately, it is in any case not very likely that it will be introduced in the near future.

Notholirion Wall. ex Boiss.

Wall. ex Boiss., *Fl. Orient.* 5: 190. 1882.

Bulb narrowly ovoid or sub-cylindric, composed of the thickened, equitant bases of the basal leaves, with a dark brown, membraneous tunic on the exterior; roots fibrous and relatively numerous, producing bulblets on their upper part; bulblets ovoid, several to a few dozen, with a rather hard outer covering when mature and several white, fleshy scales inside. Stem 20–150cm tall, glabrous. Leaves basal and cauline, the latter scattered, linear or linear-lanceolate, sessile. Inflorescence a raceme of 2–24 flowers; bracts linear; pedicels short, slightly recurved; perianth campanulate, pale purple, bluish-purple or red to pink; perianth segments 6, separate, stamens 6, filaments filiform, anthers dorsally attached at their centre, versatile; ovary cylindric or oblong; style long and slender, stigma trifid, branches subulate, slightly revolute. Capsule oblong or obovoid-oblong, bluntly angular, impressed at the apex. Seeds numerous, flattened, narrowly winged.

Altogether 4 species, distributed through China, Nepal, India, Sikkim, Bhutan, Sri Lanka and Burma. China has 3 species, occurring in the south-west and north-west.

Key to the species of the genus *Notholirion*

1. Plant 60–150cm tall; raceme with 10–24 flowers; perianth segments green at the tip; stem leaves lorate or linear-lanceolate, 10–20cm long, 1–2.5cm wide.
    2. Flowers pale purple or bluish-purple; perianth segments 2.5–3.8cm long; stamens more or less as long as the perianth segments (Tibet, Yunnan, Sichuan, Shaanxi, Gansu) .... 1. *N. bulbuliferum* (Lingelsh.) Stearn
    2. Flowers red, dark red or mauve to reddish-purple; perianth segments 3.5–5cm long; stamens slightly shorter than the perianth segments (Tibet, Yunnan, Sichuan) .... 2. *N. campanulatum* Cotton et Stearn
1. Plant 18–30cm tall; raceme with (1–) 2–4 (–7) flowers; perianth segments not green at the tip; stem leaves linear, 6–15cm long, 4–8mm wide (Tibet, Yunnan, Sichuan) .... 3. *N. macrophyllum* (D. Don) Boiss.

1. **Notholirion bulbuliferum** (Lingelsh.) Stearn in *Kew Bull.* 421. 1950. – *Paradisea bulbulifera* Lingelsh. apud Limpr. f. in *Repert. Sp. Nov. Fedde Beih.* 12: 316. 1922. – *Lilium hyacinthinum* Wilson, *Lil. East. As.* 100, t. 15. 1925. – *Notholirion hyacinthinum* (Wilson) Stapf in *Kew Bull.* 96. 1934.

Bulblets numerous, ovoid, 3–5mm in diameter, light brown. Stem 60–150cm tall, sub-glabrous. Basal leaves several, lorate, 10–25cm long, 1.5–2cm wide; stem leaves linear-lanceolate, 10–18cm long, 1–2cm wide. Raceme with 10–24 flowers; bracts leaf-like, linear, 2–7.5cm long, 3–4mm wide; pedicels slightly decurved, 5–7mm long; perianth pale purple or bluish-purple; perianth segments obovate or oblanceolate, 2.5–3.8cm long, 0.8–1.2cm wide, green at the tip; stamens more or less as long as the perianth segments; ovary pale purple, 1–1.5cm long, stigma trifid, branches slightly revolute. Capsule oblong or obovate-oblong, 1.6–2cm long, 1.5cm broad, bluntly angular.

Flowering period: July. Fruiting period: August.

Occurs in Tibet, Yunnan, Sichuan, Shaanxi and Gansu. It grows in alpine meadows or in thickets, at 3000–4500 metres above sea level. It is also distributed in Nepal, Sikkim, Bhutan and India.

† This is the most widespread of the species of *Notholirion*, occurring from southern Shaanxi and Gansu through western Sichuan into Yunnan and Tibet, and along the Himalayan range westwards into Nepal. In Shaanxi it has been collected on Tai Bai Shan, the main peak of the Qinling range, and in Yunnan its southern limit is in the Cang Shan near Dali, where it was collected in 1981 by the Sino-British Expedition. Bulbs from this collection are now growing in the peat beds at the Royal Botanic Garden, Edinburgh, and flowered in 1984 and 1985. It was probably first brought into cultivation by Purdom or Farrer, from collections made in southern Gansu in 1914–15, and is known to have flowered in Britain in 1918. But until the 1920s it was much confused with other species of the genus, particularly the then much better known *N. thomsonianum* (Royle) Stapf from the western Himalaya, so its early history in gardens is hard to untangle. E. H. Wilson was familiar with it as a wild plant in western Sichuan, where he found it to be common 'in the upland meadows around Tachien-lu' (Kangding), flowering in July 'along with Primulas, Gentians, Pedicularis in variety, *Meconopsis integrifolia* Franch. and dwarf alpine species of Rhododendron.' He says that he sent seeds 'on several occasions to England and to America', but that as far as he knew they did not germinate. Most of the plants now in cultivation probably derive from the collections of Ludlow and Sheriff made in south-east Tibet, Bhutan and Sikkim in the 1930s and 1940s, though there have been other introductions.

Notholirion is a neglected genus in gardens. None of the species could be considered easy to grow, but they are no harder than many other lilies. Their monocarpic habit no doubt reduces their popularity. It may take six or seven years for bulblets to reach flowering size. But they are certainly worth attempting, for they are very beautiful plants. This species and *N. campanulatum* produce stems three or four feet tall, with as many as 20 flowers or more, and look very fine associated with rhododendrons. Fresh seed usually germinates quite well, though sometimes needing two winters to break dormancy, and with patience it is possible to build up a good stock of plants. Their growing requirements are similar to those of *Nomocharis* and other Sino-Himalayan alpine lilies. They may sometimes need protection against late spring frosts.

2. **Notholirion campanulatum** Cotton et Stearn in *Lily Year Book* 3: 19, f. 6. 1934.

Bulblets numerous, ovoid, 5–6mm in diameter, light brown. Stem 60–100cm tall, sub-glabrous. Basal leaves numerous, lorate, 22–24cm long, 2–2.5cm wide, membraneous; stem leaves linear-lanceolate, 10–20cm long, 1–2.5cm wide, membraneous. Raceme with 10–16 flowers; bracts leaf-like, linear-lanceolate, 3–7cm long, 4–9mm wide, green; pedicels slightly decurved, 4–7mm long, 2–3mm broad; perianth campanulate, red, dark red or pink to reddish-purple, nodding; perianth segments obovate-lanceolate, 3.5–5cm long, 1–2cm wide, green at the tip; stamens slightly shorter than the perianth segments; ovary cylindric, 1–1.3cm long, 2–3mm broad; style *c.* 2cm long, stigma trifid, branches subulate, divergent. Capsule oblong, 2–2.5cm long, 1.6–1.8cm broad, light brown.

Flowering period: June–August. Fruiting period: September.

Occurs in Yunnan (north-west part), Sichuan and Tibet. It grows on grassy slopes or at the edges of mixed woodland, at 2800–3900 metres above sea level. It is also distributed in Sri Lanka and Burma.

† This is a very fine lily, with slightly larger flowers than the preceding species, which in good colour forms are of a rich shade of wine-red. It is rarer both in the wild and in gardens than *N. bulbuliferum*, but has perhaps tended to be overlooked because it was not recognized as distinct until 1934. Apart from the flower-colour (often not obvious in dried specimens), the two species can be distinguished by the relative lengths of the perianth segments and stamens. In *N. bulbuliferum* they are roughly equal, while in *N. campanulatum* the stamens are noticeably shorter. Their distribution ranges overlap, both species occurring in Yunnan, Sichuan and Tibet, but *N. campanulatum* extends into the mountains of Burma and has an outlying station in the highlands of Sri Lanka, and is not found in the eastern Himalaya. It was also collected in the Cang Shan in 1981, and has since flowered in the Royal Botanic Garden, Edinburgh. Its cultural requirements are like those of *N. bulbuliferum*.

3. **Notholirion macrophyllum** (D. Don) Boiss., *Fl. Orient.* 5: 190. 1822; Woodc. et Stearn, *Lil. World* 379, f. 125, 136, 1950. – *Fritillaria macrophylla* D. Don, *Prodr. Fl. Nepal.* 51. 1825. – *Lilium macrophyllum* (D. Don) Voss, *Vilmorins Blumeng.* 1: 1105. 1895.

Stem 20–35cm tall, glabrous. Basal leaves lorate, stem leaves 5–10, linear, 6–15cm long, 4–8mm wide. Raceme with 2–6 flowers; bracts leaf-like, narrowly linear, 1.2–2.5cm long, curved near the tip; pedicels 0.6–2cm long, slightly decurved; perianth trumpet-shaped, pale purple-red or purple, perianth segments oblanceolate-oblong, 2.5–5cm long, 0.6–1.5cm wide, apex blunt or rounded, base angustate; filaments filiform, glabrous, 2–3.5cm long, anthers long-ellipsoid, *c.* 5cm long; ovary oblong, 7–8mm long, *c.* 4mm broad; style 1.5–3.2cm long, stigma trifid, branches subulate, slightly revolute.

Flowering period: August.

Occurs in Sichuan, Tibet and Yunnan (north-east and north-west parts). It grows on grassy slopes and in woodland glades, at 2800–3400 metres above sea level. It is also distributed in Nepal and Sikkim. The whole plant can be used medicinally.

† *N. macrophyllum* is the smallest species of its genus, growing to little more than a foot tall. Its distribution extends from western Nepal eastwards as far as north-east Yunnan, and it appears to favour rather drier and sunnier situations than the two other Chinese species. It is not particularly common in any part of its range, and although, for example, it was collected by Hooker in Sikkim in 1849, it was not found there again for many decades. It is also rare in gardens. Its flowers are held more horizontally and open more widely at the mouth than in the two preceding species, and although its smaller stature makes it less impressive it is still very beautiful and desirable.

Brief Bibliography

This list is by no means exhaustive, and aims only to draw the attention of interested readers to the most important works on the subject. Those wishing to pursue research in depth will find plenty of references in the texts listed.

Monographs in Chinese

CHEN Haozi. *Hua Jing*. Beijing: Nongye, 1962 [originally published 1688].

CHEN Jingyi. *Chuan Fang Bei Zu*. Beijing: Nongye, 1982 [originally published *c*. 1255].

FANG Wenpei. *Ladingwen Zhiwuxue Mingci ji Shuyu*. Chengdu: Sichuan Renmin, 1980.

HAN E. *Si Shi Zuan Yao*. Beijing: Nongye, 1981 [originally published *c*. 950].

LI Shizhen. *Bencao Gang Mu*. Beijing: Renmin Weisheng, 1975–81 [originally published 1590].

TANG Shenwei (Ed.). *Chongxiu Zheng He Jingshi Zhenglei Beiyong Bencao*. Beijing: Renmin Weisheng, 1957 [photographic reprint of 1249 edition].

WANG Fa-tauan and TANG Tsin (Eds.). *Flora Reipublicae Popularis Sinicae/Zhongguo Zhiwu Zhi*, vol. 14: Liliaceae (1); Beijing: Kexue, 1980. (referred to in this book as *Flora RPS*).

WANG Hao, and others (Ed.). *Guang Qun Fang Pu*. Beijing: Pei Wen Zhai, 1708.

WANG Xiangjin. *Qun Fang Pu*. Yangzhou: Yu Yang Quan Ji, *c*. 1700 [originally published *c*. 1621].

XU Guangqi. *Nong Zheng Quan Shu*. Shanghai: Shanghai Guji, 1979 [originally published 1639].

ZHAO Xuemin. *Bencao Gang Mu Shi Yi*. Beijing: Renmin Weisheng, 1963 [originally published 1765].

Monographs in English

COX, E. H. M. *Plant-hunting in China*. London: Collins, 1945.

ELWES, H. J. *Monograph of the Genus Lilium*. London: 1877–80.

EVANS, A. *The Peat Garden and its Plants*. London: Dent, 1974.
FOX, D. *Growing Lilies*. London: Croom Helm, 1985.
GROVE, A. and COTTON, A. D. *Supplements to Elwes' Monograph of the Genus Lilium*. Parts 1–7. London: 1933–40.
SYNGE, P. M. *Lilies*. London: Batsford, 1980.
TURRILL, W. B. *Supplements to Elwes' Monograph of the Genus Lilium*. Parts 8 & 9. London: 1960–62.
WILSON, E. H. *The Lilies of Eastern Asia*. London: Dulau, 1925.
WOODCOCK, H. D. and STEARN, W. T. *Lilies of the World*. London: Country Life, 1950.

Periodical articles

BALFOUR, I. B. The genus *Nomocharis*; in *Transactions of the Botanical Society of Edinburgh* 27: 273–300. 1918.
BARANOVA, M. V. K sistematika roda *Lilium* L.; in *Novosti Sistematiki Vysshikh Rastenii* 8: 89–95. 1971.
BARANOVA, M. V. On the dwarf lily *Lilium pumilum*; in *Botanicheskii Zhurnal* 56 (6): 787–97. 1971.
COMBER, H. F. *Lilium cathayanum*; in *Journal of the Royal Horticultural Society* 70: 78–9, ff. 27, 28. 1945.
COMBER, H. F. A new classification of the genus *Lilium*; in *Lily Year Book (RHS)*, 1949: 86–105.
NODA, S. Chromosomes of diploid and triploid forms found in natural populations of tiger lily in Tsushima; in *Botanical Magazine (Tokyo)* 91 (1024): 279–83. 1978.
SEALY, J. R. *Nomocharis* and *Lilium*; in *Kew Bulletin*, 1950: 273–97.
SEALY, J. R. A revision of the genus *Nomocharis* Franchet; in *Botanical Journal of the Linnean Society* 87: 285–323. 1983.
SEN, S. Intraspecific differentiation in karyotype of *Lilium*; in *Cytologia (Japan)* 43 (2): 305–15. 1978.
SMITH, W. W. Notes on Chinese lilies; in *Transactions of the Botanical Society of Edinburgh* 28: 122–60. 1922.
STAPF, O. *Lilium, Notholirion* and *Fritillaria*; in *Kew Bulletin*, 1834; 94–6.

Descriptions of new sections of the genus Lilium

Lilium sect. **Asteridium** S. G. Haw, sect. nov. – *Lilium sect. Pseudolirium* Wils., *Lil. E. Asia*, 50 (1925), p.p.; *Lilium sect. Sinomartagon* Comber, *Lily Yrbk.* 13, 101 (1949), p.p.; *Lilium sect. Lophophorum* Wang et Tang, *Flora Reipub. Pop. Sin.*, Tomus 14, 118 (1980), p.p.

Bulbus albus, parvus; radices caulinae evolutae. Caulis usque ad 90cm altus, saepe brevior, glaber vel pubescens. Folia sparsa, sessilia, linearia vel linearo-lanceolata. Flores erecti; perianthium apertum, stellare; segmenta perianthii non vel apicem versus leviter recurvata, basi non vel vix in unguem attenuata. Cotyledones epigeae.

Typus sectionis: **Lilium concolor** Salisb.

Bulb white, small; stem roots present. Stem up to 90cm tall, often shorter, smooth or pubescent. Leaves scattered, sessile, linear or linear-lanceolate. Flowers erect; perianth open, starry; perianth segments not or slightly recurved towards the apex, not or scarcely narrowed into a claw at the base. Cotyledons epigeal.

As can be seen from the synonymy, *Lilium concolor* has been placed by previous authors in several different and very distinct sections, and no consensus of opinion about its relationships has emerged. It is so difficult to associate with any of the other species of *Lilium* that it seems best to accord it its own section. It may be most closely related to the lilies of section Pseudolirium, but differs from them in having perianth segments which are not or scarcely narrowed to a claw at the base; in its foliation, which shows no tendency towards the formation of a whorl just below the flowers; and in its epigeal germination. Its erect flowers and scattered, linear leaves separate it from lilies of other sections.

Lilium sect. **Dimorphophyllum** S. G. Haw, sect. nov. – *Lilium sect. Sinomartagon*

Comber, *Lily Yrbk.* 13, 101 (1949), p.p.; *Lilium sect. Martagon* Wils., *Lil. E. Asia*, 61 (1925), p.p.

Caulis 40–200cm altus, glaber. Folia sparsa, breviter petiolata, manifeste dimorpha, inferiora late vel anguste lanceolata, superiora multo breviora, ovata vel sub-orbiculata, bracteis similia. Flores nutantes; segmenta perianthii valde recurvata. Cotyledones epigeae.

Typus sectionis: **Lilium henryi** Baker.

Stem 40–200cm tall, smooth. Leaves scattered, shortly petiolate, clearly dimorphic, the lower ones broadly or narrowly lanceolate, the upper ones much shorter, ovate or sub-orbiculate, similar to the bracts. Flowers nodding; perianth segments strongly recurved. Cotyledons epigeal.

The description of this section is necessarily rather abbreviated because one of its two species, *Lilium rosthornii*, is imperfectly known. The dimorphous leaves set this group apart from all other lilies, with the possible exception of *Lilium wardii*, which seems sometimes to show similar foliation. There is no overwhelming reason to associate them with the lilies of section Sinomartagon, as has been usual in the past; the similarity in flower-form is insufficient basis for the assumption of such a relationship. *Lilium henryi* has been crossed both with *Lilium speciosum* and lilies of section Regalia, and is probably closely related to both.

Societies of interest to lily enthusiasts

Listed below are some of the main lily societies throughout the world, and the names and addresses of their secretaries at the time of writing (1986).

Australia:

The Australian Lilium Society:
Mr J. H. Young, 24 Halwyn Crescent, West Preston, Victoria, Australia 3072

Canada:

The Canadian Prairie Lily Society:
Dr E. A. Maginnes, University of Saskatchewan, Saskatoon, Saskatchewan

New Zealand:

The New Zealand Lily Society:
Mr J. Gover, PO Box 1394, Christchurch

UK

The Royal Horticultural Society Lily Group:
Mrs A. C. Dadd, 21 Elmbrook Rd, Wokingham, Berks. RG11 1HF

USA:

The North American Lily Society Inc.:
Mrs D. B. Schaefer, PO Box 476, Waukee, Iowa 50263

Most of these societies arranges sales of surplus bulbs and/or seed exchanges.

Other societies operating useful seed exchanges include those with an interest in alpine plants. One recent seed list of the Alpine Garden Society, for example, included no less than 44 species of *Lilium* and seven of *Nomocharis* and *Notholirion*. Almost half of these were Chinese species. The most useful societies to join, with the names and addresses of their secretaries, are given on page 162.

UK:

The Alpine Garden Society:
Mr E. M. Upward, Lye End Link, St. Johns, Woking, Surrey GU21 1SW

The Scottish Rock Garden Club:
Miss K. M. Gibb, 21 Merchiston Park, Edinburgh EH10 4PW

USA:

The American Rock Garden Society:
Mr N. Singer, Norfolk Rd., S. Sandisfield, Mass. 01255

Some recently available information

Since the completion of the main text of this book, Miss Liang Sung-yun has informed me of the results of her latest researches. A number of new species have recently been described in Chinese botanical publications, and considerable work has been undertaken on the genus *Nomocharis*. In order to make this book as up to date as possible, this appendix has been added, containing a summary of the latest Chinese work.

In the genus *Lilium*, two completely new species have been described, and for two others a change of status has been suggested. The new species are:

Lilium medogense S. Y. Liang, *Acta Phytotaxonomica Sinica* 23 (5): 392–393 (1985).

This is described as being similar to *L. paradoxum* Stearn, but differing in having larger flowers of yellow colour, with elliptical perianth segments, 5–6cm long, 2–2.4cm wide, and slightly saccate at the base. The type specimen was collected from *Abies* forest near Medog in south-eastern Tibet on 26 June 1980 (W. L. Chen no. 10625, in the Herbarium of the Botanical Institute of Academia Sinica, Beijing).

This is extremely interesting as being the only lily similar to the unusual *L. paradoxum* ever to be found. It would appear to be so like *L. paradoxum* that its status as a distinct species might be doubted, especially as so little material of either lily exists in Herbaria. Further collections are needed to elucidate the relationship and precise status of these two lilies.

Lilium habaense Wang et Tang, *Acta Botanica Yunnanica* 8 (1): 51–52 (1986).

This lily is said to be similar to *L. fargesii* Franch., but to differ in having a smooth stem, smooth nectaries and the style shorter than the ovary. The type specimen was collected on 8 June 1939, from open rocks of the Haba range on the Zhongdian plateau (K. M. Feng no. 1229, at the Botanical Institutes in Beijing and Kunming). As far as I can tell from the description and illustration, this lily differs in no obvious respects from *L. stewartianum* Balf. f. et W. W. Sm., and I strongly suspect that it belongs with that species.

In the same paper which contains the description of *L. habaense*, Miss Liang alters the status of *L. nanum* Klotz. et Garcke var. *brevistylum* S. Y. Liang, making it a distinct species, *L. brevistylum* S. Y. Liang, justifying this on the grounds that it has a purple bulb, purplish-yellow flower, shorter style and shorter filaments.

Miss Liang has also undertaken a considerable amount of work on the pollen of Chinese lilies. The result of this with regard to the genus *Nomocharis* will be discussed below, but in the genus *Lilium* it has led her to reduce *L. amoenum* Wilson to a subspecies of *L. sempervivoideum* Lévl. It thus becomes *L. sempervivoideum* Lévl. subsp. *amoenum* (Wilson ex Sealy) Liang. These two lilies are certainly very closely related, and this treatment should probably be accepted.

There have been some very interesting developments in the taxonomy of the genus *Nomocharis*. Firstly, a new species has been described; a complete translation of the original description follows:

Nomocharis biluoensis Liang, *Bulletin of Botanical Research/Zhiwu Yanjiu* 4 (3): 169–170 (1984).

Close to *N. synaptica* Sealy, from which it differs in having dissimilar inner perianth segments, one being larger, broadly ovate and acute, the others ovate-lanceolate, and in having oblong anthers, 6–7mm long, and the style shorter than the ovary.

Bulb ovoid, 3–4cm tall, *c.* 4cm in diameter, scales numerous, erect, imbricate, fleshy, lanceolate. Stem 80cm tall, smooth. Leaves scattered, lanceolate, 8–10cm long, 1.2–1.5cm wide. Flowers 6–8, borne singly on spreading pedicels 7.5–11cm long, perigon *c.* 7–9.5cm in diameter, white, densely and irregularly spotted with dark purple; outer tepals ovate-lanceolate, acuminate, base sub-rotundate, 3.5–5cm long, 1.5–1.8cm wide, entire, inner tepals dissimilar, one broadly ovate, acute, 3.5–4.5cm long, 2.8cm wide, the other two ovate-lanceolate, acuminate, at base sub-rotundate, 3.5–5cm long, 1.6–2.4cm wide, with flabellate, fleshy cristate protrusions, 6mm long, 8mm wide, on each side of the basal nectarial furrow; stamens 1.5cm long; filaments cylindric, narrowing towards the top, 1–1.3cm long, purple, fleshy, abruptly contracted into a pale yellow awn, 3mm long; anthers oblong, 6–7mm long, purple, dorsifixed, versatile. Ovary cylindric, dark purple, 8–10mm long, 2mm broad; style shorter than the ovary, fine, expanded towards the apex, 6mm long; stigma capitate, trilobed.

Yunnan: Weixi, in *Abies* forest, alt. 3400m, 13 July 1981. Hengduan Mountains Expedition of the Institute of Botany 1485. (Type, in the Herbarium of the Botanical Institute of Academia Sinica, Beijing.)

This plant would appear to be very similar indeed to *N. synaptica*, and confirms my suspicion that such plants occur within the borders of China. Very probably the plants identified as *N. synaptica* which turned up in cultivation in Scotland did indeed derive from collections made in Yunnan, and may be closer to this plant than to Kingdon Ward's Assam collections. Whether *N. biluoensis* can really be considered distinct from *N. synaptica*, and whether either should be separated from *N. aperta*, still in my opinion remain open questions. Miss Liang now accepts that both *N. saluenensis* Balf. f. and *N. aperta* (Franch.) Wilson properly belong in this genus, and not in *Lilium*, as they were treated in the *Flora RPS*. But her studies of the pollen of *Nomocharis* species have led her to draw a clear distinction between *N. saluenensis* and *N. aperta* on the one hand, and all the other species of the genus, including her *N. forrestii* and *N. biluoensis*, on the other. This is a remarkable conclusion, which does not accord with what would have been expected from examination of the gross morphology of these plants. The pollen examined, however, came from no more than two specimens of each of these species (from only one specimen in the case of *N. biluoensis*, of which only a single specimen appears to exist), and it must be suspected that a wider range of specimens might show less clear differentiation. The *N. aperta-N. forrestii-N. synaptica-N. biluoensis* group is undoubtedly very complex and difficult to classify satisfactorily. At present I still favour the view that a single, variable species should be recognized, possibly with several subspecies, but it may be that there really are three or four distinct species involved. Factors such as hybridization could account for the intermediate forms which seem to link these species. Extensive studies of the variation of natural populations of these plants are needed before any final conclusions can be reached.

The Chinese work summarized above was originally published in the following periodical articles:

LIANG, S. Y. A new species of *Lilium* from Xizang; in *Acta Phytotaxonomica Sinica* 23 (5): 392–393. 1985.

WANG Fatauan, *et al.* Notes on Chinese Liliaceae XII; in *Acta Botanica Yunnanica* 8 (1): 51–52. 1986.

LIANG, S. Y. and ZHANG, W. X. Pollen tetrads in *Lilium* with a discussion on the delimitation between *L. sempervivoideum* and its ally; in *Acta Phytotaxonomica Sinica* 22 (4): 297–300. 1984.

LIANG, S. Y. Studies on the genus *Nomocharis* (Liliaceae); in *Bulletin of Botanical Research/Zhiwu Yanjiu* 4 (3): 163–178. 1984.

LIANG, S. Y. and ZHANG, W. X. Pollen morphology of the genus *Nomocharis* and its delimitation with *Lilium*; in *Acta Phytotaxonomica Sinica* 23 (6): 405–407. 1985.

I am greatly indebted to Miss Liang Sung-yun for drawing my attention to these articles and supplying me with copies of them.

Index

Major references to species, sections etc. are in bold type. Valid species are shown in roman *type, synonyms in italics.*